Design of
Supports in Mines

Design of Supports in Mines

CEMAL BIRÖN

ERGIN ARIOĞLU

Department of Mining Engineering
Istanbul Technical University

A Wiley-Interscience Publication

JOHN WILEY & SONS

New York • Chichester • Brisbane • Toronto • Singapore

Copyright © 1983 by John Wiley & Sons, Inc.

All rights reserved. Published simultaneously in Canada.

Reproduction or translation of any part of this work
beyond that permitted by Section 107 or 108 of the
1976 United States Copyright Act without the permission
of the copyright owner is unlawful. Requests for
permission or further information should be addressed to
the Permissions Department, John Wiley & Sons, Inc.

Library of Congress Cataloging in Publication Data:

Birön, Cemal.
　　Design of supports in mines.

　　"A Wiley-Interscience publication."
　　Bibliography: p.
　　Includes index.
　　1. Ground control (Mining)　I. Arıoğlu, Ergin.
II. Title.
TN288.B57　1982　　　622'.28　　　82-15921
ISBN 0-471-86726-8

Printed in the United States of America

10　9　8　7　6　5　4　3　2　1

Foreword

Dr. Cemal Birön, Chairman of the Department of Mining Engineering, Istanbul Technical University, and a Visiting Professor of the Department of Mining and Minerals Engineering, Virginia Polytechnic Institute and State University, during the 1980–1981 academic year, has taught three courses for undergraduate and graduate students entitled "Principles of Rock Mechanics," "Introduction to Mining Engineering," and "Coal Mine Ground Control." The notes and handouts of the last course have been incorporated in this book. The mining literature contains treatises on rock mechanics and several on mine supports, but a need has been identified in the area concerning the design and calculation of the actual dimensions of mining supports. The resulting book by Dr. Birön and Dr. Arıoğlu meets this need well by reinforcement with numerical examples for a variety of roof-support systems. This substantial contribution may be used as a textbook for students in mining and mineral engineering and a resource and reference book for practicing and professional mining engineers for their working libraries.

It is my pleasure to endorse this creative effort, which makes a substantial contribution to the literature of mining engineering by being a useful, applied treatment of this important area.

J. RICHARD LUCAS
Head, Department of Mining and Minerals Engineering
Virginia Polytechnic Institute and State University
Blacksburg, Virginia

December 1982

Preface

The course "Coal Mine Ground Control" induced us to write *Design of Supports in Mines*. It was written to give the "design concept" to mining students. Chapters cover wood supports, gallery steel arches, roof bolts, steel longwall supports, concrete supports, and stowing as a support. Each chapter describes the physical, mechanical, and engineering properties of the materials used, such as wood, steel, concrete, and stowing. The pressures involved in galleries, longwalls, and so on are calculated using practical formulas. The designs of gallery wooden sets, steel arches, roof bolts, props-and-caps, powered supports, shotcrete, concrete shaft lining, hydraulic filling, and the like are given as numerical examples with actual dimensions.

CEMAL BIRÖN
ERGIN ARIOĞLU

Istanbul, Turkey
January 1983

Acknowledgments

The authors acknowledge the following publishers for granting permission to make use of materials published by them: Birsen Publishing House, Istanbul, Springer-Verlag, Berlin, Dunod, Paris, Pergamon Press, Oxford, and John Wiley & Sons, New York. We thank architects M. Sahin and N. Arıoglu for preparing the cover drawing; Miss P. Adkins for typing the English text; and Mrs. T. Riggan, Miss B. Hutchinson, and A. Yüksel for producing the drawings for this book. We extend our gratitude to Dr. J. Richard Lucas and Dr. E. Topuz of Virginia Polytechnic Institute and State University for their encouragement. Finally, the authors take this opportunity to thank E. W. Smethurst and C. Mikulak of John Wiley & Sons for their unfailing care in the publication of this book.

C.B.
E.A.

Contents

INTRODUCTION   1

1. WOODEN SUPPORTS   3

   1.1 **Status of Wood Supports in Mines, 3**
   1.2 **Engineering Characteristics of Mine Timber, 4**
      1.2.1 Fibrous Structure, 4
      1.2.2 Factors Affecting Wood, 4
      1.2.3 Strength of Timber, 5
   1.3 **Pressure on Wooden Supports, 27**
      1.3.1 Evaluation of Pressures, 27
      1.3.2 Pressures on Roadways, 28
      1.3.3 Pressure in Longwalls, 32
   1.4 **Design of Wooden Supports, 39**
      1.4.1 Design Principles, 39
      1.4.2 Wooden Gallery Sets, 39
      1.4.3 Additions to Gallery Sets, 49
      1.4.4 Optimum Design, 52
      1.4.5 Design of Longwall Supports, 55

2. STEEL GALLERY SUPPORTS   61

   2.1 **Importance of Steel, 61**
   2.2 **Engineering Characteristics of Steel, 62**
      2.2.1 Chemical Structure, 62
      2.2.2 Mechanical Characteristics, 62
      2.2.3 Characteristics of Support Elements, 65

2.3 Design of Rigid Arches, 70

    2.3.1 Description of Rigid Arches, 70
    2.3.2 Stress Evaluation, 70
    2.3.3 Design of Arch Profile, 73
    2.3.4 Numerical Application, 74

2.4 Design of Articulated (Moll) Arches, 76

    2.4.1 Description of Articulated Arches, 76
    2.4.2 Design of a Moll Arch with Two
          Articulations, 76
    2.4.3 Design of a Moll Arch with Three
          Articulations, 81

2.5 Design of Yielding Arches, 82

    2.5.1 Description of Yielding Arches, 82
    2.5.2 Estimation of Yielding Arches, 86

3. ROOF BOLTS AND TRUSSES   89

3.1 Principle of Roof Bolts, 89
3.2 Varieties of Roof Bolts, 90

    3.2.1 Slot-and-Wedge Bolts, 90
    3.2.2 Expansion-shell Roof Bolts, 92
    3.2.3 Grouted Roof Bolts, 95
    3.2.4 Resin Roof Bolts, 96
    3.2.5 Wooden Roof Bolts, 101
    3.2.6 Testing of Roof Bolts, 101

3.3 Design of Roof Bolts, 105

    3.3.1 Stability of Bolted Blocks, 105
    3.3.2 Length of Bolts, 107
    3.3.3 Spacing of Bolts, 107
    3.3.4 Diameter of Bolts, 108
    3.3.5 Density of Bolts, 108
    3.3.6 Numerical Example, 109

3.4 Application of Roof Bolts, 111

    3.4.1 Room Entries, 111
    3.4.2 Gateways of Longwalls, 112
    3.4.3 Longwall Faces, 114
    3.4.4 Metal Ore Mines, 116

3.5 Advantages of Roof Bolting, 118
3.6 Roof Trusses, 119
 3.6.1 Principle and History of Roof Trusses, 119
 3.6.2 Design of Roof Trusses, 120

4. **STEEL LONGWALL SUPPORTS**   **125**

4.1 Evolution of Steel Longwall Supports, 125
4.2 Steel Props and Caps, 128
 4.2.1 Friction Props, 128
 4.2.2 Hydraulic Props, 130
 4.2.3 Articulated Caps, 130
4.3 Design of Props and Caps, 134
 4.3.1 Prop Density Calculations, 134
 4.3.2 Intrusion of Props, 136
 4.3.3 Size of Caps, 136
4.4 Powered Supports, 138
 4.4.1 Development of Powered Supports, 138
 4.4.2 Types of Powered Supports, 138
 4.4.3 Description of Powered Supports, 144
4.5 Design of Powered Supports, 147
 4.5.1 Dimensions Related to Supporting, 147
 4.5.2 German System, 149
 4.5.3 English System, 150
 4.5.4 Austrian Systems, 153
 4.5.5 French System, 155
 4.5.6 Polish System, 157
 4.5.7 American System, 162
4.6 Advantages and Disadvantages of Powered Supports, 165
 4.6.1 Advantages of Powered Supports, 165
 4.6.2 Disadvantages of Powered Supports, 166
4.7 Applicability of Powered Supports, 167
 4.7.1 Roof Conditions, 167
 4.7.2 Floor Conditions, 167
 4.7.3 Seam Thickness, 168
 4.7.4 Seam Inclination, 168

4.7.5 Small Faults, 168
4.7.6 Water at the Face, 168
4.7.7 Life of the Panel, 168
4.7.8 Face Length Rate of Advance, 169
4.7.9 Number of Shifts, 169
4.7.10 Hints for Good Installation, 169

5. **CONCRETE SUPPORTS**   **171**

5.1 Importance of Concrete, 171
    5.1.1 Advantages of Concrete, 171
    5.1.2 Disadvantages of Concrete, 172
5.2 Constituents of Concrete, 172
    5.2.1 Cement, 173
    5.2.2 Aggregates, 173
    5.2.3 Other Constituents, 174
    5.2.4 Water, 175
5.3 Engineering Characteristics of Concrete, 175
    5.3.1 Water–Cement Ratio, 175
    5.3.2 Compaction, 177
    5.3.3 Granulometry of Aggregates, 178
    5.3.4 Curing Conditions, 178
    5.3.5 Working Conditions, 179
    5.3.6 Making of Concrete, 180
    5.3.7 Transportation of Concrete, 181
    5.3.8 Pouring and Maintenance of Concrete, 182
    5.3.9 Strength of Concrete, 183
5.4 Applications of Concrete in Mines, 184
    5.4.1 Shotcreting, 184
    5.4.2 Monolithic Concreting, 188
    5.4.3 Gallery Lining with Concrete Blocks, 188
    5.4.4 Concrete Shaft Lining, 189
    5.4.5 Artificial Roofs, 192
5.5 Design Concrete, 192
    5.5.1 Design in Concrete Preparation, 192
    5.5.2 Design for Shotcreting, 197
    5.5.3 Design of Shaft Lining, 198

5.5.4 Design of Shaft Indentation, 200
5.5.5 Design of Artificial Roofs, 202
Appendix 5.1 Relevant British Standards, 203
Appendix 5.2 Selected List of Relevant ASTM
Standards, 205

6. STOWING   209

6.1 Importance of Stowing, 209
6.1.1 Amount of Stowing Materials, 209
6.1.2 Sources of Stowing Materials, 210
6.1.3 Advantages of Stowing, 212
6.1.4 Disadvantages of Stowing, 213
6.2 Application of Stowing Systems, 213
6.2.1 Hand Stowing, 213
6.2.2 Gravity Stowing, 215
6.2.3 Mechanical Stowing, 215
6.2.4 Pneumatic Stowing, 215
6.2.5 Hydraulic Stowing, 219
6.2.6 Consolidated Stowing, 222
6.3 Design of Hydraulic Stowing, 223
6.4 Economics of Stowing, 230

References   233

Index   239

Notation

A_y	Side reaction
a	Distance between sets
B_y	Side reaction
b	Spacing of bolts
CvT	Convergence at the face
c	Distance between wedges
D	Prop density
D_{max}	The biggest aggregate size
d	Diameter of the bolt, diameter of the pipe
d_b	Diameter of the cap
d_e	Diameter of cut on the cap at the corner
d_y	Diameter of the post
E	Elasticity modulus, expansion rate of immediate roof
F	Cross-section area of the cap, cross-section area of the profile, cross-section area of the bolt, minimum capacity of power support
F_t	Area of anchorage
f	Factor related to wood material, Protodyakonov coefficient of rock hardness
H	Depth
h	Load height, height of immediate roof, height of cross section
h'	Vertical distance of the arch
J	Unit head loss in pipes
K	Statistical constant, factor of expansion of immediate roof, final convergence, fineness modulus, coefficient for volumetric concentration of mixture
K_f	Coefficient according to the floor rock
K_t	Coefficient according to the support of gateway rib

K'	Heaving of the floor
L	Roof index
L_b	Length of the cap
L_k	Length of the buckling
L_y	Length of the post, length, span
M	Bending moment, total turning moment
M_{max}	Maximum bending moment
m	Thickness of seam
m_d	Thickness of stowing
m_0	Bolt density
N	Normal force
n	Factor of safety
OMS	Output per manshift
P	Anchorage force, side pressure on lining
P_k	Breaking load
P_t	Total load produced by parabolic dome
p	Porosity
Q_k	Amount of stowing
Q_w	Amount of water
q	Load per unit length
q_t	Roof load per unit length
R	Reaction, force
R_{max}	Carrying capacity of bolt
r	Radius of arch, radius of shaft
S	Standard deviation
T	Maximum shear force, tensioning load
t	Thickness of lining
V	Volume, variation coefficient, velocity of mixture
Y	Closure of sides
W	Section modulus
α	Loading factor, water-cement ratio in weight
β	Ratio where the central post is placed
γ	Density
δ	Angle of inclination
Λ	Factor of completeness, compaction
λ	Number of slenderness, friction coefficient of pipes
μ	Coefficient of friction between roof rock and bolt steel
σ_b	Bending stress

σ_c	Compressive strength
σ_f	Stress in floor
σ_k	Tensile breaking strength
σ_n	Normal stress
σ_{sf}	Allowable bending stress, allowable stress in steel
σ_t	Roof pressure
σ_y	Side pressure
τ_{max}	Maximum shear stress
τ_s	Shearing stress
τ_{sf}	Allowable shear stress
φ	Angle of friction of rocks, Durand coefficient
ω	Factor of buckling

Design of
Supports in Mines

Introduction

The design of supports in mines is a basic requisition for the mining engineer—the first step toward effective roof control. The elements of the design procedures are quite involved so only the high points are considered in this book.

Wood supports, although outdated in some countries, are still a basic support material for many areas where steel cannot be used. The strength of mine timber is studied under tensile, crushing, bending, and shear stress conditions and approximate engineering data given. Then, pressure acting on wooden gallery sets are longwall supports are evaluated. The design of wooden gallery and longwall set elements such as caps, posts, wedges, and auxilliary supports, are calculated. Emphasis is on the economical factors, that is, designing with the least amount of material.

Steel has now taken the place of wood in many mines. The engineering characteristics of steel such as stresses, hardness, mode of failure, moment of inertia, ranking ratio and design procedure of rigid arches, articulated (Moll) arches, and yielding arches are considered. Special emphasis is given to roof bolting and the design of roof bolts of various types.

Steel longwall supports, such as friction props and hydraulic props, are discussed in detail and special emphasis is given to powered supports and their design, including several European practices.

The engineering characteristics of concrete support (especially shotcreting) are studied in depth, and the design of shaft and gallery lining is given.

The last support system discussed is stowing. Hydraulic stowing, which is coming into more effective use, is dealt with in detail and the required design data are concisely set out.

CHAPTER 1

Wooden Supports

1.1 STATUS OF WOOD SUPPORTS IN MINES

Wood (timber) was the most important material for support in mining operations until the end of the second world war. Since then steel has become the primary material used for mine supports. The reason for considering wood as a support material is that it is still in use in small-scale coal and metal mines.

Timber is a light-weight material, easily transported and easily manipulated in supporting systems. Oak wood has a density of 0.73 g/cm^3 and a bending strength of 1200 kg/cm^2. It is 11 times lighter but 2 times weaker than steel. This makes timber an economical material when used in "short-lifetime" supports.

Wood has both advantages and disadvantages when used in mines. Although their importance is reduced, it is still sought in many mining operations support systems. The advantages are as follows:

1. It is light, easily carried, cut, manipulated, and put in the form of a mine support.
2. It breaks along definite fibrous structures, giving visual and audible signs before it fails completely. This has given wood a psychological advantage for miners over steel.
3. Broken pieces can be reused for wedges, fillings, and so on.

The disadvantages are as follows:

1. The mechanical strengths (bending, tensional, buckling, sheer, compression) are dependent on fibrous structures and natural defects occurring within the wood.

2. Humidity has a very pronounced effect on the strength.
3. Many fungi, living in humid conditions, affect the timber, considerably diminishing its strength.
4. Timber is an easily combustible material, and fires may spread quickly in supports and product poisonous gases.

1.2   ENGINEERING CHARACTERISTICS OF MINE TIMBER

1.2.1   Fibrous Structure

Wood is composed of about 45–50% cellulose, 20–25% lignin, 5% pectin, and 20% other materials [1, p. 56]. Cellulose is a polysaccharide forming the walls of wood cells. These cells are called fibers. Lignin is the cementing substance in wood. It is a three-dimensional polymer of phenylprophane units. Pectin is starchy jelly material that binds cell walls and is highly susceptible to swelling and shrinking as a result of exposure to water [2, p. 315].

The structure of the wood is shown in Fig. 1.1. The "living" cells form a thin layer on the outside just beneath the bark of the tree. From year to year this layer dies off and forms "layers of age," the "hard" part of the wood, the essential section of the timber.

1.2.2   Factors Affecting Wood

Water.   Water is the most important component in timber. About 25% of the water content is in the living cell, with the remaining 75% in the voids of fibers. A newly cut tree contains 35–50% water. The loss of water in the voids is due to temperature and relative humidity of the environment. At normal conditions (20°C and 80% relative

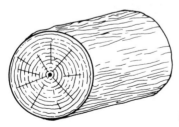

Figure 1.1   Macroscopic structure of wood.

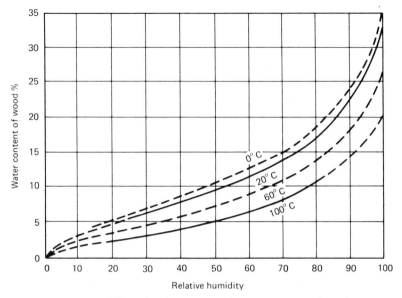

Figure 1.2 Effect of environment on the water content of wood.

humidity), the water content is about 20%. Any wood having less than this amount of water is considered *dry*, while wood having more than 30% water is considered *wet*. The effect of environment is seen in Fig. 1.2 [2, p. 317].

Defects of Timber. As a natural material, wood has many defects caused by growing conditions. Knots, the bases of tree branches, affect bending strength. In addition, growth rings may not be concentric owing to wind and sun conditions, and the drying conditions may form cracks. These defects are illustrated in Fig. 1.3 [2, p. 319].

1.2.3 Strength of Timber

Mine timber is subject to bending, compression, buckling, and shear. The strength of timber under these conditions and the factors affecting such strengths are given in the following sections.

Tensile Strength. Wood's greatest strength is tensile strength, especially parallel to the fibrous structure. The tensile strengths of some materials are shown in Table 1.1 [1, p. 323].

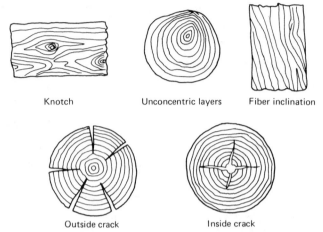

Knotch   Unconcentric layers   Fiber inclination

Outside crack   Inside crack

Figure 1.3   Natural structural defects of wood.

Table 1.1   Tensile Strengths of Some Materials[a]

Material	Tensile Strength (kg/cm^2)
Steel wire, max	32,000
Iron wire, hard drawn	5500–8400
Steel, building	5200–6200
Copper wire, hard drawn	4200–4900
Rayon, acetate	10,000
Silk	3500
Cotton	2800–8000
Hemp	8800–9000
Coniferous woods	500–1500
Broad-leaved woods	200–2600
Bamboo	1000–2300

[a]See reference 1.

The tensile strength of wood parallel to the grain* is extremely high and may reach, for some species in air-dried conditions ($u = 12\%$), a maximum of 3000 kg/cm². Many factors affect such values. The values of Jayne [4], for several wood species for early and late wood

*"Grain" and "fiber" are used interchangeably in referring to the fibrous structure of wood.

Table 1.2 Average Tensile Strength of Wood Fibers[a]

Wood Species	Earlywood Fibers (kg/cm^2)	Latewood Fibers (kg/cm^2)
Redwood	4850	9140
Sitka spruce	8230	9070
Slash pine	3300	6470
Douglas fir	3590	9980
White fir	5130	7310
Red cedar	3340	4780
White pine	4220	4640

[a]See references 4 and 1.

fibers are given in Table 1.2 [1, p. 322]. Unfortunately, this high tensile strength cannot be utilized in construction for several reasons.

The relationship of loading direction to the grain angle has a pronounced effect on tensile strength. Baumann has studied the variations for some woods for tensile, bending, and compression strengths, and the deviations are shown in Fig. 1.4 [1, p. 326].

Density has a positive relation to tensile strength, as shown on Finnish pine (Fig. 1.5) [1, p. 327].

On the contrary, moisture decreases tensile strength. Many investigators have pointed out that, from a 10% moisture content to the

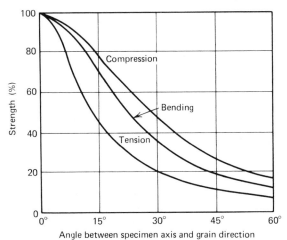

Figure 1.4 Relation between tensile strength and grain direction [1].

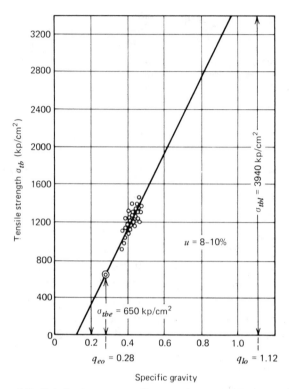

Figure 1.5   Relation between tensile strength and specific gravity [1].

fiber saturation point, there is a linear decrease of tensile strength, as shown in Fig. 1.6. According to the U.S. Forest Products Laboratory, an increase in moisture content of 1% lowers the tensile strength along the grain about 3%. According to many investigators, the peak tensile strength is found when the moisture content is between 8 and 10% [1, p. 327].

Knots and crotches also reduce the strength of wood, as the grain is distorted in passing around them; the fibers of knots are nearly at a right angle to the fibers of the wood.

Crushing Strength.   The maximum crushing strength plays an important role in the utilization of wood. For air-dry woods the maximum crushing strength parallel to the grain reaches, on the average, only about 50% of the tensile strength along the grain. The different behavior of wood in compression and tension may be explained by its

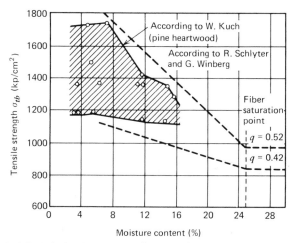

Figure 1.6 Relation between tensile strength and moisture content [1].

fibrous structure. The tightly wedged and cemented fibers sustain very high tensile stresses; in compression, probably an early buckling of individual fibers occurs, starting failure [1, p. 335].

The effect of loading angle and grain direction is more pronounced in crushing strength than with tensile strength. Kollmann's [5] investigations on pine and beech woods, are seen in Fig. 1.7 and 1.8 [1, p. 341–342].

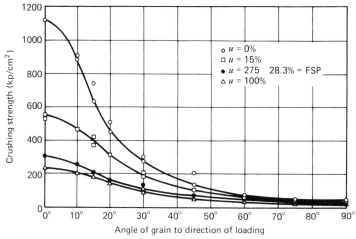

Figure 1.7 Relation between crushing strength of pine wood and angle of grain [1,5].

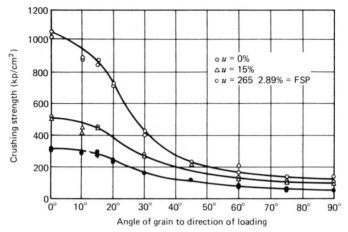

Figure 1.8 Relation between crushing strength of beech wood and angle of grain [1, 5].

The compressive strength of wood along the grain increases with density, not only for single species but also for the total range of densities of all species. Fig. 1.9 illustrates this for oven-dry, air-dry and water-saturated conditions [1, p. 345].

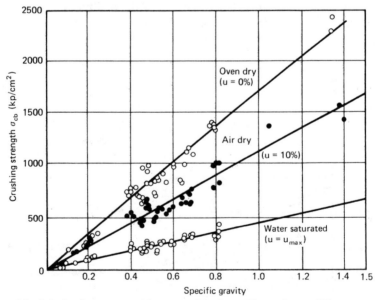

Figure 1.9 Relation between crushing strength and specific gravity at different water contents [1].

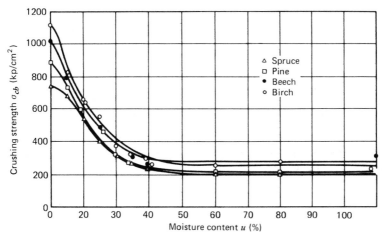

Figure 1.10 Relation between crushing strength and moisture content for many species [1, 5].

Moisture content is the most important factor in the crushing strength of wood. As water is deposited between the micelles, it causes reduction of the intercellar attractive forces, and therefore cohesion. This effect is seen in the investigations of Kollmann [5] illustrated in Fig. 1.10 [1, p. 348]. After moisture content reaches about 18%, the strength does not differ much, being reduced to about half of that under dry conditions.

The investigations of Dixon and Hogan [6] at the S.M.R.S agree well with these findings, as seen in Fig. 1.11. After a moisture level of

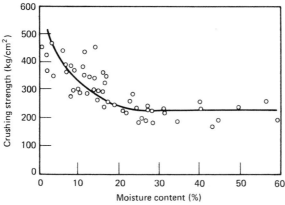

Figure 1.11 Effect of moisture on the crushing strength of wood [6].

20% is reached, the crushing strength falls from 500 kg/cm² to about 250 kg/cm² and does not decrease any more. The recent investigations of Saxena and Singh [7] at the Central Mining Research Center of India also indicate the prominent effect of moisture in decreasing the crushing strength in wood [2, 12]. As seen in Figs. 1.11 and 1.12 the load bearing capacity of wooden props falls significantly with increasing moisture content. After 15% moisture the effect becomes unimportant.

The effect of knots and crotches on crushing strength is not as great as on tensile strength, nevertheless, this problem should not be underestimated [1, p. 353].

Buckling Strength.   Buckling strength is measured parallel to the fibers, on the axis of the timber. If the ratio of length to diameter is less than 11, the crushing strength in compression is utilized [1, p. 414]. According to investigators, the buckling strength of wood [2, p. 329] depends on the following:

$$\sigma = \frac{\pi^2 E}{\lambda^2} \qquad\qquad \text{for } \lambda > 100 \qquad\qquad (1.1)$$

$$\sigma = \sigma_c(1 - a\lambda + b\lambda^2) \qquad \text{for } \lambda < 100 \qquad\qquad (1.2)$$

where λ = slenderness ratio = $4l/d$
        E = elasticity modulus of timber

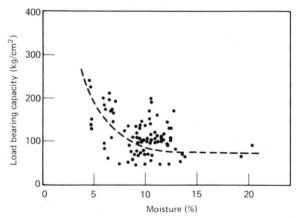

Figure 1.12   Moisture versus load bearing capacity of wood [7].

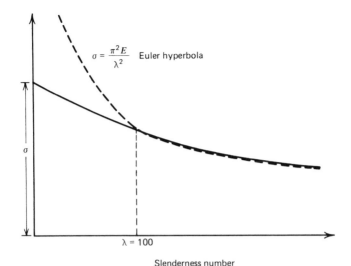

Figure 1.13 Buckling strength versus slenderness [2].

σ = buckling strength of timber
σ_c = crushing strength of timber
a, b = quality constants of timber; for regular mine timber $a = 0$, $b = 2$
l = length of timber
d = diameter of timber

Table 1.3 gives the buckling strength for regular mine timbers.
Saxena and Singh [7] give the following buckling strength formula and values that resulted from many mine timber tests conducted in India. See Fig. 1.14 [7, p. 12].

$$P = 47.2 - 1.5h/d \qquad (1.3)$$

where P = load bearing capacity of prop in tonnes
h = height of the prop, in millimeters
d = average diameter of the prop, in millimeters

Bending Strength (Modulus of Rupture). Horizontal timbers are subject to conditions of bending stress where the upper fibers are under compression and the lower fibers under tension. The neutral

Table 1.3 Buckling Strength of Mine Timber[a]

Diameter [d(cm)]	Length [l(m)]	Slenderness $\lambda = 4l/d$	Buckling Strength (kg/cm²)	Density (g/cm³)	Water Content (%)
16.1	1.00	24.8	284.0	0.560	20.3
13.2	1.00	30.3	384.3	0.616	21.5
12.7	1.20	37.8	322.1	0.637	19.4
12.5	1.20	38.4	221.7	0.555	23.2
14.2	1.50	42.3	280.7	0.686	21.9
16.5	1.50	36.4	207.0	0.585	24.3
16.5	1.80	43.7	175.2	0.670	25.7
16.0	1.80	45.0	271.3	0.630	22.2
13.5	2.00	59.0	214.5	0.638	21.3
16.1	2.00	49.0	233.9	0.664	24.3

[a]See reference 2.

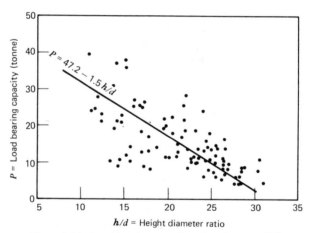

Figure 1.14 Load bearing capacity of a mine prop [7].

Table 1.4 Load Bearing Capacity of Props[a]

Height-diameter ratio (h/d)	5	10	15	20	25
Load bearing capacity (tonnes)	39.7	32.2	24.7	17.2	9.7

[a]See reference 7.

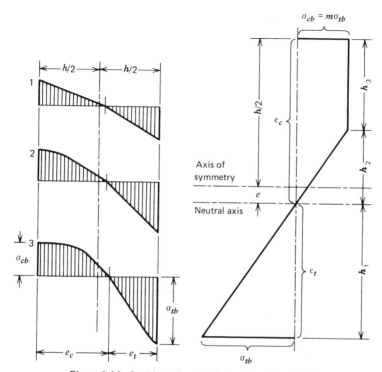

Figure 1.15 Position of neutral axis of bending [1, 8].

axis lies closer to the side of tension than to compression because the tensile strength is much greater than its compressive strength, as illustrated in Fig. 1.15 [8; 1, p. 361].

The modulus of rupture is measured by loading on the center of a beam as shown in Fig. 1.16 [2, p. 330]. On loading, the deflection is measured and plotted as illustrated. There are several zones of deformations. The first is the elastic zone where load and deflection are proportional. In the second zone this relationship continues, although less in degree. Finally, at the peak load of P_K, the outermost fiber breaks. The breaking is not sudden but stretches from fiber to fiber, as shown in the lower illustration. This essential behavior of timber gives visual and audio indication of breaking while still carrying some load and giving enough time to change the support in the mines. The psychological effect on miners should be evaluated.

The bending strength or modulus of rupture is calculated as follows:

$$\sigma_b = \frac{M_{max}}{W} \tag{1.4}$$

$$M_{max} = \frac{P_k l}{4} \tag{1.5}$$

$$W = \frac{bh^2}{6} \tag{1.6}$$

$$\sigma_b = \frac{P_k l/4}{bh^2/6} = \frac{3}{2}\frac{P_k l}{bh^2} \tag{1.7}$$

where σ_b = bending strength (modulus of rupture)
$\quad M_{max}$ = maximum bending movement
$\quad P_k$ = breaking load
$\quad l$ = span, length of beam
$\quad W$ = section modulus
$\quad b$ = breadth of beam
$\quad h$ = height of beam
$\quad f$ = deflection
$\quad \Lambda$ = work done by deflection

The work done by deflection can be measured by the area under the curve or by Λ (shaded area, Fig. 1.16) as follows:

$$\Lambda = \int P \, df \tag{1.8}$$

The maximum work done by the load making deflection is $P_k f_{max}$. The ratio n of the work to the maximum work is the "factor of completeness" and is a measure of the quality of the wood [1, p. 365].

$$n = \frac{\Lambda}{P_k f_{max}} = \frac{\text{shaded area}}{P_k f_{max}} \tag{1.9}$$

The value of n is about 0.7 but may fall to 0.5 owing to natural defects such as knots and crotches.

The direction or orientation of the fiber in relation to the load affects the bending stress as well. The work of Winter [9] for ashwood in Fig. 1.17 shows tensile, bending, and crushing strengths. The crushing strength is least affected by the fiber direction [1, p. 366].

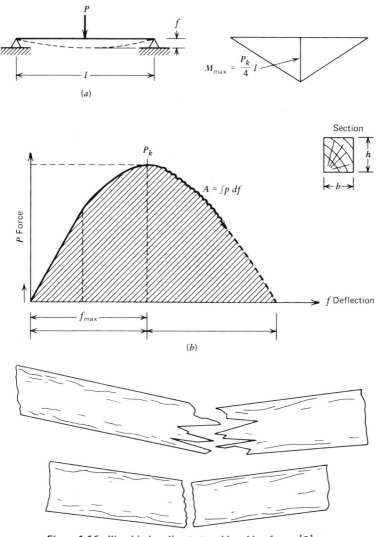

Figure 1.16 Wood in bending test and breaking forms [2].

The effects of moisture and temperature, linear decrease [1, p. 369], are shown in Fig. 1.18 [10].

The effects of knots and crotches are given in the work of Siimes [11] on the woods at Finland. Knots reduce the modulus of rupture of wood considerably if they are located in the tension zone near the

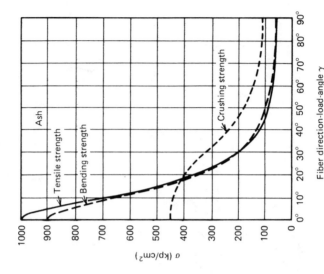

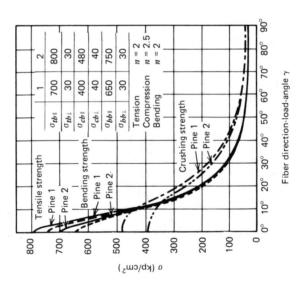

Figure 1.17 Effect of fiber direction on the tensile, crushing, and bending strengths of wood [1, 11].

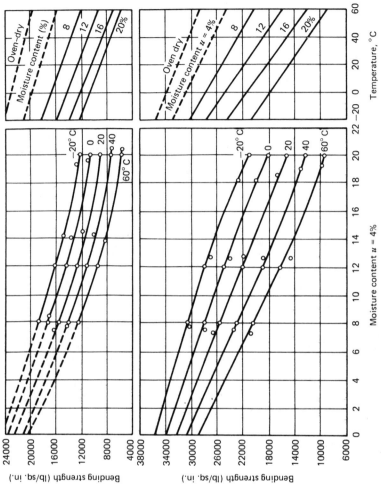

Figure 1.18 Effect of moisture and temperature on the bending strength of wood [1, 10].

19

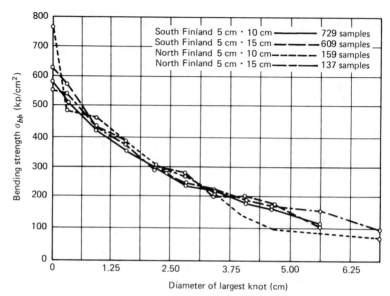

Figure 1.19 Effect of the diameter of knots on the bending strength of wood [1, 11].

critical cross section [1, p. 373]. The diameter of the knots reduces the bending strength considerably. Figure 1.19 clearly indicates this effect.

Duration of stresses (fatigue) is progressive in its effect on wood. According to the work of Graf [12], as shown in Fig. 1.20, the load bearing capacity of wood falls to 60% after 10–20 days [1, p. 376].

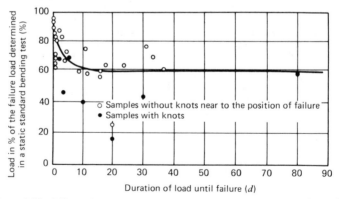

Figure 1.20 Effect of duration of load on bending strength of wood [1, 12].

Shear Strength. The ultimate shearing strength of wood is remarkably lower than the torsional strength. According to the *Wood Handbook* [13, p. 82], for solid wood members, the allowable ultimate torsion shear may be taken as the shear stress parallel to the grain, and two-thirds of this value may be used as the allowable torsional shear stress at the proportional limit. The shearing stress perpendicular to the grain, is about three to four times higher than that parallel to grain [1, p. 414].

Newlin and Wilson [14] found a parabolic function between shearing strength τ_s and density R, where weight is oven dry based on green volume as follows:

$$\tau_s = AR^{4/3} \tag{1.10}$$

where τ_s = shearing strength, in kilograms per square centimeter
$\quad R$ = density (volume oven dry divided by green volume), in grams per cubic centimeter
$\quad A$ = tangential parallel to grain conditions:
$\quad\quad$ 193 green
$\quad\quad$ 281 air dry ($u = 12\%$)

$\quad\quad$ radial parallel to grain conditions:
$\quad\quad$ 179 green
$\quad\quad$ 255 air dry ($u = 12\%$)

Moisture, as always, affects the strength. The work of Ehrmann [15] is seen in Fig. 1.21 [1, p. 404].

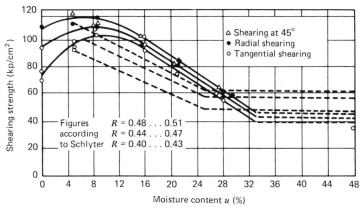

Figure 1.21 Effect of moisture content on the shearing strength of wood [1, 15].

Numerical Values of Strengths. Table 1.5, based on the work of Keylwerth [15], using prismatic specimens, lists many mechanical properties of some commercial woods, across the grain [1, p. 333]. Table 1.6 [1, p. 353] gives the crushing strengths of some woods, according to work by Graf [12]. Table 1.7 [1, p. 396] shows the torsional strengths of wood, based upon data from many investigators.

The safe allowable stresses in wood constructions are given in Table 1.8 [2, p. 336] based upon the standards of the Turkish Association of Bridges and Construction [16].

Allowable Strengths. Because wood is a natural material, many unknown factors affect its strength and necessitate a large "safety factor." The best practice would be to test the strength of the material in hand and use it according to these determined strengths. The "safer stresses" can be calculated by the following formula:

$$\sigma_{sf} = \frac{\overline{X} - KS}{n} f_k f_y \qquad (1.11)$$

where σ_{sf} = safe (allowable) stress

$\overline{X}$ = average strength obtained on small specimens without any defect

K = statistical constant ensuring a small probability of the strength being exceeded; in general, K is taken as 2.

S = standard deviations obtained on small specimens without any defect

n = safety factor for many cases of loading of supports for long duration. For loading in bending $n = 2.25$; for crushing and shearing $n = 1.4$.

f_k = a factor for natural defects. In English standards f_k is 0.40–0.75. A prop full of knots and cracks is taken as 0.5. Props should be stocked according to their defects.

f_y = a factor for duration of loading. For long durations, $f_y = 1$; for short duration, $f_y = 1.5$.

A numerical example clarifies this formula. Let us assume that in the testing on oak props of no natural defects the average bending strength $\overline{X}$ is 1270 kg/cm² and standard deviations S is 300 kg/cm².

Table 1.5 Mechanical Properties of Some Commercial Woods[a]

Species	Tensile Strength σ_{tb} (kp/cm²)	Bending Strength σ_{bb} (kp/cm²)	Shock Resistance a (kp·cm·cm²)	Cleavage $\sigma_{cl} = \dfrac{P_{max}}{F}$ kp/cm²		Density q_u (g/cm³)	Moisture Content u (%)
				Cleavage Specimen	Double Cleavage Specimen		
Oak	83–97 / 90	128–141 / 135	1.2–1.4 / 1.3	8.4–9.5 / 8.8	20.2–31.8 / 26.3	0.64–0.68 / 0.66	12.1–12.6 / 12.4
Beech	96–118 / 107	135–146 / 142	1.4–1.6 / 1.5	8.3–8.8 / 8.6	24.6–38.1 / 31.1	0.67–0.73 / 0.69	11.5–11.8 / 11.7
Hornbeam	– / –	58–91 / 80	0.9–1.3 / 1.2	7.1–9.7 / 8.1	27.1–33.3 / 30.2	0.76–0.78 / 0.77	10.8–11.3 / 11.2
Ash	101–127 / 112	62–167 / 113	1.3–1.8 / 1.5	8.5–10.0 / 9.6	25.4–42.3 / 30.6	0.67–0.79 / 0.76	8.6–9.1 / 0.8
Walnut	98–114 / 105	169–183 / 175	1.6–2.2 / 1.9	8.8–12.7 / 11.7	21.1–32.6 / 29.7	0.58–0.66 / 0.60	10.9–11.4 / 11.2
Lime	50–60 / 58	90–92 / 91	1.0–1.7 / 1.3	6.6–7.3 / 7.0	20.7–25.4 / 22.8	0.57–0.59 / 0.58	9.8–10.1 / 10.0
Alder	69–79 / 73	92–99 / 96	0.9–1.4 / 1.2	6.4–7.4 / 7.0	24.5–26.5 / 25.9	0.53–0.56 / 0.55	10.4–10.8 / 10.7
Spruce	33–40 / 38	42–49 / 46	0.7–1.0 / 0.9	4.1–5.1 / 4.6	15.0–20.4 / 17.3	0.48–0.69 / 0.54	11.5–11.6 / 11.7
Larch	48–52 / 50	69–77 / 75	0.9–1.1 / 1.0	4.8–5.5 / 5.2	11.3–27.9 / 20.1	0.66–0.72 / 0.68	11.6–12.3 / 12.1
Pine	32–37 / 34	53–64 / 57	0.7–1.0 / 0.8	4.7–5.4 / 5.1	15.1–19.6 / 17.4	0.51–0.58 / 0.55	10.9–13.7 / 12.0

[a]See references 1 and 15.

23

Table 1.6 Effect of Knots on Crushing Strength[a]

| Species | Specific Gravity q_0 | Moisture Content (%) | Dimensions of the Specimen | | | Spiral Grain (Divergence of Fibers, cm per 1 m Length) | Maximum Diameter of Knots on One Surface (cm) | Crushing Strength (kp/cm^2) |
			a (cm)	b (cm)	h (cm)			
Pine	0.48	>25	6.54	10.09	34.91	9	Without knots	185
Pine	0.48	>25	6.32	10.54	28.80	6	2.2	113
Pine	0.56	>25	6.29	10.06	49.72	10	3.3	105
Pine	0.60	>25	6.36	10.38	49.30	20	5.0	95
Pine	0.41	~14	6.42	7.54	39.48	8	Without knots	329
Pine	0.48	~14	7.53	11.44	34.90	11	2.8	279
Pine	0.46	~14	6.39	7.51	49.57	8	3.0	235
Spruce	0.43	~14	5.92	11.51	29.60	0	Without knots	354
Spruce	0.40	~14	7.00	12.10	50.16	8	1.4	292
Spruce	0.42	~14	7.97	10.95	49.80	9	2.9	263

[a]See references 1 and 16.

Table 1.7 Torsional Strength of Woods[a]

Species	Density q (g/cm^3)	Moisture Content u (%)	Torsional Strength (kp/cm^2) Sticks with Fibers Parallel to the Longitudinal Axis	Torsional Strength (kp/cm^2) Sticks with Fibers Perpendicular to the Longitudinal Axis	Deformation at Failure (°/cm) Sticks with Fibers Parallel to the Longitudinal Axis	Deformation at Failure (°/cm) Sticks with Fibers Perpendicular to the Longitudinal Axis	Source
Coniferous Woods							
Spruce	0.35–0.42	11.9	87–101	32–36	0.96–1.33	1.8–2.0	K. Huber
Spruce	0.45	12	$\{$ $\tau_1 = 186$ / $\tau_2 = 181$	25	–	–	H. Carrington
Spruce	0.48	10.8	141–162	30–62	1.1	2.3	O. Kraemer
Pine	0.50–0.55	12.2	134–140	43–50	0.95–1.8	1.37–1.42	K. Huber
Pine, heartwood	0.65	10.8	163–178	39–64	0.7	1.0	O. Kraemer
Pine, sapwood	0.56	11.3	135–167	–	0.6–0.9	–	O. Kraemer
Broadleaved Species							
Birch	0.67	12	200[1]	–	–	–	O. Kraemer
Beech	0.66–0.69	11.3	246–250	151–156	2.6–3.0	2.0	K. Huber
Oak	0.67–0.71	11.7	190–220	112–114	1.2–2.0	1.0–1.5	K. Huber
Ash	–	9	158–213–250	–	1.5–2.5	–	F. Kollmann
Ash	–	12.0–16.3	140–188–238	–	1.6–3.0	–	F. Kollmann
Ash	–	air dry	210–245	–	–	–	R. Baumann
Ash	0.81–0.83	11.3	258–277	156–158	1.8–2.1	1.2–1.5	K. Huber
Ash	0.65	9.5	179–262–345	136–167–209	1.3	1.3–1.7	O. Kraemer
Walnut	0.60	8.1	275–303–326	135–152–162	1.0–1.2	1.4	O. Kraemer

[a]See reference 7.

Table 1.8 Safe Stresses in Wood Constructions (kg/cm^2)^a

	Class 1		Class 2		Class 3	
Type of Stress	Pine	Oak	Pine	Oak	Pine	Oak
Bending	130	140	100	110	70	75
Bending in continuous beams	140	155	110	120	75	80
Tension parallel to grain	105	110	85	100	0	0
Crushing parallel to grain	110	120	85	100	60	70
Crushing perpendicular to grain	20	30	20	30	20	30
Shear parallel to grain	9	12	9	10	9	10
Shear perpendicular to grain	27	36	27	30	27	30

^{a}See reference 2.

Let us find the safe stress for bending for defective props used for long duration. In formula (1.11) $K \cong 2$.

$$\sigma_{sf} = \frac{\bar{X} - KS}{n} f_k$$

$$= \frac{1270 - 2 \times 300}{2.25} \times 0.50$$

$$\simeq 150 \text{ kg/cm}^2$$

If we do not make this investigation and rely instead on safe stresses given in Table 1.8, the stress allowed would be 75 kg/cm^2 for third-class oak wood under bending. Then economy demonstrated of $150/75 = 2$ (two times) is quite important in engineering. The following safety factor is used:

$$\text{safety factor} = \frac{\text{average strength measured}}{\text{safe stress taken}}$$

$$= \frac{1270}{150} = 8.47$$

This is quite large due to unknown defects. In stress calculations a safety factor 2–4 is used.

1.3 PRESSURES ON WOODEN SUPPORTS

1.3.1 Evaluation of Pressures

There are two principles in designing wooden supports:

1. The supports should carry the loads "safely" (safety factor).
2. The amount of material and work should be held to a minimum (economy factor).

The engineer should do his best to meet the requirements of these two principles. Usually, the material needed is not economical. The engineer, in making tests on the material, may adopt higher safe stresses. After making measurements on loads in the mines, the engineer may assume lower pressure and finally a more economical design, depending upon his experience and good judgment. The steps taken in calculations may be summarized as illustrated in Fig. 1.22 [2, p. 377].

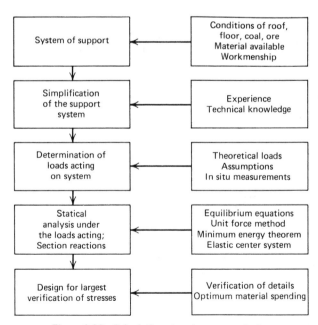

Figure 1.22 Calculation steps in support design.

1.3.2   Pressures on Roadways

According to many investigators, the pressure on a gallery is in the form of a parabolic dome [17, p. 680]. As the theoretical formulas are very complicated, for practical purposes the approximate values are accepted. The Protodyakonov formula, as shown in Fig. 1.23 [2, p. 382], is as follows:

$$h = \frac{l}{f} \tag{1.12}$$

$$f = \frac{\sigma_c}{100} \tag{1.13}$$

$$\sigma_t = \gamma h \tag{1.14}$$

$$q_t = \sigma_t a \tag{1.15}$$

$$P_t = \tfrac{4}{3}\, lha\gamma \tag{1.16}$$

where h = height of the parabola m as load height

l = half of gallery width m may be taken as length of cap on the wooden set

f = Protodyakonov coefficient of hardness can be taken from Table 1.9 or as 0.01 of the compressive strength of the

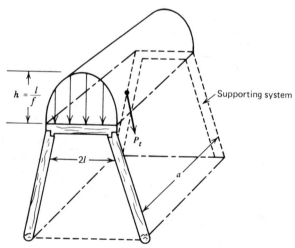

Figure 1.23   Loads on wooden gallery set according to Protodyakonov [2, 17].

Table 1.9 Protodyakonov Hardness Coefficients and Internal Angle of Friction of Rocks[a]

Rock Formations	f	φ
Quartzite, basalt, hardest rocks	20	87°08'
Hard granite, hard sandstone	15	86°11'
Quartzite veins, marble, hard gneiss-dolomite	10	84°18'
Hard limestone soft granite, marble, gneiss, dolomite	8	83°31'
Ordinary sandstone, iron ore	6	80°32'
Sandy shale, shaly sandstone	5	75°41'
Shaly schist, soft sandstone-limestone and conglomerate	4	75°58'
Weak schist, hard marl	3	71°34'
Soft schist, very soft limestone, salt rock, frozen soil, marl, broken sandstone, stony soil	2	63°26'
Gravels, broken schist, soft conglomerate, hard coal, hard shale	1.5	56°19'
Hard shale, coal	1	45°
Light sandy clay	0.8	38°40'
Peat, sandy clay, wet sand	0.6	30°58'
Sand, fine gravel, broken soil, broken coal	0.5	26°35'
Mud, other earth	0.3	16°42'

[a]See reference 17.

rock in which the gallery is driven; it is a dimensionless number

σ_c = compressive strength of rock, in kilograms per square centimeter

γ = density of rock in tonnes per cubic meter (t/m^3)

σ_t = pressure on the support in tonnes per square meter (t/m^2)

q_t = load per unit length in tonnes per meter (t/m)

a = distance between wooden sets in meters

P_t = total load produced by parabolic dome in tonnes (t).

As a numerical application of the Protodyakonov formula, let us calculate the unit load, the total load carried by a wooden support set 1.8 m wide spaced at 1-m intervals, driven in marl of 300 kg/cm^2 compressive strength at 2.5 t/m^3 density. Comprehensive strength, densities, and other structural characteristics of rocks are given in Table 1.10.

Table 1.10 Physical and Structural Characteristics of Some Rocks[a]

Rocks	Density, Dry (t/m³)	Young's Modulus (1000 × kg/cm²)	Poisson's Number	Porosity (%)	Compressive Strength (kg/cm²)	Tensile Strength (kg/cm²)	Bending Strength (kg/cm²)	Shear Strength (kg/cm²)
Batholitic Rocks								
Granite–Grandiorite	2.5–2.75	300–700	5–8	0.1–2	1200–2800	40–70	100–200	50–80
Gabbro	2.92–3.05	600–1000	5–8	2–5	1500–2000	50–80	100–220	40–85
Extrusive Rocks								
Riolite	2.45–2.60	100–200	5–10	0.4–4	800–1600	50–90	100–220	40–110
Dacite	2.50–2.75	80–180	5–11	0.5–5	800–1600	30–80	90–200	30–100
Andesite	2.30–2.75	120–350	5–9	0.2–8	400–3200	50–110	130–250	50–120
Basalt	2.75–3.00	200–1000	5–7	0.2–1.5	300–4200	60–120	140–260	50–130
Diabase	2.90–3.10	300–900	5–8	0.3–0.7	1200–2500	60–130	120–260	60–100
Volcanic tuff	1.30–2.20	–	5–10	8–35	50–600	5–45	30–80	10–40
Sedimentary Rocks								
Sandstone	2.10–2.50	150–170	8–15	1–8	100–1200	15–60	40–160	20–60
Limestone (fine)	2.60–2.85	500–800	5–10	0.1–0.8	500–2000	40–70	50–150	30–70
Limestone (coarse)	1.55–2.30	–	8	2–16	40–600	10–35	25–70	15–35
Limestone (freshwater)	1.55–2.50	–	8–15	1.5–6	400–2000	15–50	30–90	20–50
Dolomite	2.20–2.70	200–300	5–12	0.2–4	150–1200	25–60	40–160	25–70
Shale-clay	2.45–2.72	–	–	0.2–0.4	–	–	200–300	–
Metamorphic Rocks								
Marble	2.65–2.75	600–900	5–9	0.1–0.5	500–1800	50–80	80–120	35–80
Gneiss	2.60–2.78	250–600	5–11	1–5	800–2500	40–70	80–200	30–70

[a]See references 2 and 17.

$$l = \frac{1.8}{2} = 0.9 \text{ m}$$

$$h = \frac{l}{300/100} = \frac{0.9}{3} = 0.3 \text{ m}$$

$$\sigma_t = \gamma h = 2.5 \times 0.3 = 0.75 \text{ t/m}^2$$

$$q_t = \sigma_t a = 0.75 \times 1 = 0.75 \text{ t/m}$$

$$P_t = \tfrac{4}{3} \times 0.9 \times 0.3 \times 1 \times 2.5$$

$$= 0.9 \text{ t}$$

The Everling formula considers the load as a function of gallery width as demonstrated in Fig. 1.24 [2, p. 383].

$$h = \alpha L_a \tag{1.17}$$

$$\sigma_t = h\gamma$$

$$q_t = \sigma_t a$$

$$P_t = \alpha L_a^2 a\gamma \tag{1.18}$$

where h = height of load in meters

α = loading factor; depends upon rock formations; under normal conditions, 0.25–0.5; with a bad roof with many cracks, it may be 1.0–2.0

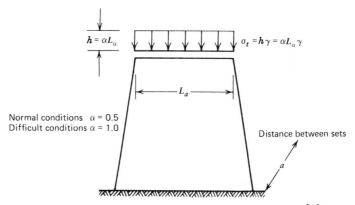

Figure 1.24   Load on wooden gallery set according to Everling [2].

As a numerical application of the Everling formula, where the width of a gallery is 1.8 m, the distance between sets is 1 m, the density of rock is 2.5 t/m^3, and $\alpha = 0.25$, we have

$$h = \alpha L_a = 0.25 \times 1.8 = 0.45 \text{ m}$$

$$\sigma_t = h\gamma = 0.45 \times 2.5 = 1.125 \text{ t/m}^2$$

$$q_t = \sigma_t a = 1.125 \times 1 = 1.125 \text{ t/m}$$

$$P_t = \alpha L_a^2 a\gamma = 0.25 \times (1.8)^2 \times 1 \times 2.5$$

$$= 2.025 \text{ t}$$

A comparison of the Protodyakonov and Everling formulas leads to these conclusions. The larger load calculated as a result of the Everling formula necessitates heavier supports and a wider margin of safety. The formula is easier to use in calculating the support dimensions. On the other hand, the Protodyakonov formula can be calculated more precisely and give better results in weak broken formations. The engineer may judge the formation conditions and accept a reasonable factor for load calculation.

The side pressure in hard rock is very small or negligible. Broken rocks exert side pressure up to twice that of the roof load.

1.3.3   Pressure in Longwalls

The cross section of a longwall face with wooden supports is shown in Fig. 1.25 [2, p. 387]. According to pressure arch theory, the main load of strata above the face of a longwall is transferred to the coal in front of the face as "front abutment". At the longwall face, there is a "destressed," or "relaxed," zone where only the load of the immediate (false) roof is left to be carried by support. If the immediate roof is very weak, it caves easily, and by expansion it fills the gob which supports the main roof. This is the working case with most caved longwalls. If the immediate roof does not cave, special attention is directed to making it cave to fill the gob. If this uncaved distance is large, the weight on the face support is heavy. In such cases "stowing" systems are used to fill the gob completely by the use of pneumatic or hydraulic systems to enable the immediate roof to sag without breaking. The pressure are calculated according to the conditions of the immediate roof.

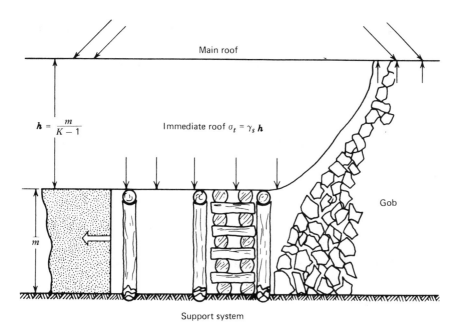

Figure 1.25 Cross section of a longwall face and height of the immediate roof showing support system [2].

In Fig. 1.25 the height of the immediate roof is given by the following formulas:

$$h = \frac{m}{K - 1} \tag{1.19}$$

$$K = 1 + E \tag{1.20}$$

$$E = \frac{\gamma_s - \gamma_k}{\gamma_k} \tag{1.21}$$

$$h = m \frac{\gamma_k}{\gamma_s - \gamma_k} \tag{1.22}$$

$$\sigma_t = h\gamma_s \tag{1.23}$$

where h = height of immediate (false) roof, in meters
K = factor of expansion of immediate roof
m = thickness of seam, in meters
E = expansion rate of immediate roof

γ_s = density of immediate roof (solid), in tonnes per cubic meter (t/m^3)

γ_k = density of immediate roof (broken), in tonnes per cubic meter (t/m^3)

σ_t = pressure of immediate roof, in tonnes per square meter (t/m^2)

The calculation of pressure in a seam 2 m thick, with a false roof density of 2.5 t/m^3 in solid and 1.8 t/m^3 in broken condition is as follows:

$$h = 2 \, \frac{1.8}{2.5 - 1.8} = 5.15 \text{ m}$$

$$\sigma_t = 5.15 \times 2.5 = 12.875 \text{ t/m}^2$$

According to Siska [19] the pressure on the support is determined by the following formula, represented in Fig. 1.26 [2, p. 341]:

$$\sigma_t = m\gamma\alpha_1\alpha_2\alpha_3 \, \frac{1}{K-1} \qquad (1.30)$$

$$\alpha_1 = \frac{V_t + V_a}{V_t} = 1 + \frac{x + 0.5h \tan\varphi}{l} \qquad (1.31)$$

$$\alpha_3 = \frac{V_1}{V_t} \qquad (1.32)$$

$$m_e = m - m_d \qquad (1.33)$$

where σ_t = roof pressure on support, in tonnes per square meter (t/m^2)

m = seam thickness, in meters

γ = density of immediate roof, in tonnes per cubic meter (t/m^3)

α_1 = factor of caving; calculated as in Eq. (1.31), Fig. 1.26a, or as given in Table 1.11

α_2 = factor of stowing; caving = 1.0; hand stowing = 0.7; pneumatic stowing = 0.5; hydraulic stowing = 0.12

α_3 = factor of self-support of immediate roof calculated as in Eq. (1.32), Fig. 1.26b, or as given in Table 1.12

K = factor of expansion

V_t = volume of immediate roof supported, in cubic meters

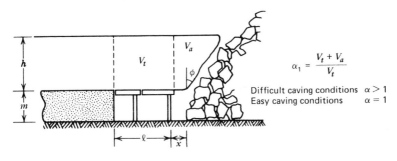

$$\alpha_1 = \frac{V_t + V_a}{V_t}$$

Difficult caving conditions $\alpha > 1$
Easy caving conditions $\alpha = 1$

(a)

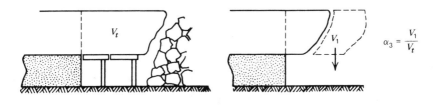

$$\alpha_3 = \frac{V_1}{V_t}$$

(b)

Figure 1.26 Pressures on supports according to Siska [2, 19].

Table 1.11 Factor of Caving α_1 According to Geometrical Configuration of Roof[a]

Roof Conditions	Geometrical Dimensions	Factor of Caving α_1
Easily caved roof rock (Category 1)	$x = 0$ $\varphi = 0°$	1.0
Regularly caved sometimes delayed (Category 2)	$x = 0.5$ m $\varphi = 40°$ $m < 1.5$ m	$\alpha_1 = 1 + \dfrac{0.5 + 2.5\ m}{l}$
Strong roof rock naturally caved with difficulty (Category 3)	$x = 1.7$ m $\varphi = 15°$ $m > 1.5$ m	$\alpha_1 = 1 + \dfrac{0.5 + 0.8\ m}{l}$
	$\varphi = 10°$ $m > 1.5$ m	$\alpha_1 = 1 + \dfrac{1.7 + 0.9\ m}{l}$
Complete stowing	$m_e = m - m_d$	$\alpha_1 = 1 + \dfrac{x + 5\ m_e \tan \varphi}{l}$

[a] See references 2 and 19.

Table 1.12   Factor of Self-Support α_3[a]

Immediate Roof Conditions	Lithology	Stowing of Gob	α_3
Easily caved	Coarse shale bands	Caving	0.75
	Fine shale bands	Pneumatic	0.40
Regularly caved sometimes delayed	Shaly silt	Caving	0.50
	Fine, medium grained sandstone	Pneumatic	0.35
Strong roof, hardly caved	Coarse band shale	Caving	0.40
	Coarse grained sandstone–conglomerate	Pneumatic	0.35

[a] See references 2 and 19.

V_a = volume of immediate roof cantilevered, in cubic meters
V_1 = volume of immediate roof in an unsupported face, in cubic meters
V_2 = volume of immediate roof supported and cantilevered, in cubic meters
l = width of face supported, in meters
x = width of face unsupported, in meters
h = height of immediate roof, in meters
φ = angle of break, degrees; taken from the vertical
m_e = relative thickness of seam, in meters
m_d = thickness of stowing, in meters

Figure 1.27 [2, p. 393] illustrates investigations of the Ostrava Research Institute [19] of pressures on various face conditions. The stresses are calculated in terms of seam thickness. For example, in easily caved roofs, the stress on a 2-m seam is about 12 t/m² in the stowing system (lower curve) and about 16 t/m² in a caved face (upper curve). The categories are given according to the breakage of drill cores and the findings can be summarized as follows:

Under static roof conditions, the roof pressure acting on supports increases with seam thickness.

Under changing roof conditions, the roof pressure acting on supports is less in stowed faces compared to caved faces.

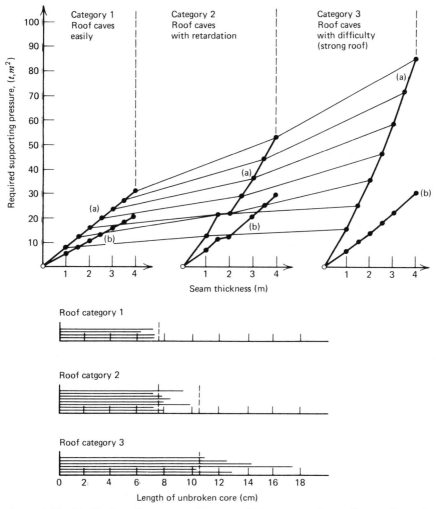

Figure 1.27 Distribution of pressures with seam thickness according to Ostrava Research Institute [2, 19] ; (*a*) caving system; (*b*) stowing system.

Under very stable roof conditions and in thick seams, stowing systems should be used to diminish the pressure.

As a numerical application, take an easily caved face of 2 m. The seam, with caving, includes coarse grained shale bands. Use the expansion factor $K = 1.35$, density 2.5 t/m^3.

$$\alpha_1 = 1, \quad \alpha_2 = 1, \quad \alpha_3 = 0.75, \quad K = 1.35$$

$$\sigma_t = 2 \text{ m} \times 2.5 \text{ t/m}^3 \times 1 \times 0.75 \times \frac{1}{1.35 - 1}$$

$$= 10.70 \text{ t/m}^2$$

This is in accordance with previous calculations of longwall pressure (12.875 t/m²).

Another method of calculating longwall pressure uses the formula of Terzaghi [20], devised for pressure calculations for tunnels at shallow depth, with conditions of roof rock as loose material. The formula adopted for longwall stresses is shown in Fig. 1.28 [2, p. 396]. This formula was successfully used in the design of a reinforced concrete roof for a thick coal seam [21]. The theory behind the formula is well explained by Evans [22].

$$\sigma_t = \frac{\gamma B}{K \tan \varphi} \tag{1.34}$$

$$B = B_1 + m \tan \left(45° - \frac{\varphi}{2} \right) \tag{1.35}$$

where σ_t = pressures on supports, in tonnes per square meter (t/m²)
γ = density of immediate roof, in tonnes per cubic meter (t/m³)
B = half of the width of face subjected to loading, in meters

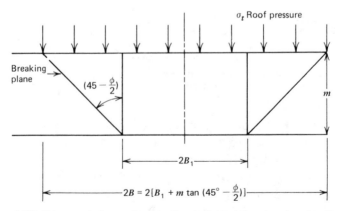

Figure 1.28 Pressures in longwalls according to Terzaghi's approximation [2, 20].

B_1 = half of the actual width of face, in meters
m = seam thickness, in meters
φ = angle of internal friction of roof rock, in degrees
K = an empirical coefficient, may be taken as $K = 1$

As the numerical application of Terzaghis formula, let us calculate the pressure on a longwall face with 2-m seam thickness, 4 m in width, a roof density of 2.5 t/m^3 and internal friction angle of 40°.

$$B = B_1 + m \tan \left(45° - \frac{40°}{2} \right)$$

$$= 2 + 2 \times 0.4663 = 2.93 \text{ m}$$

$$\sigma_t = \frac{2.5 \text{ t/m}^3 \times 2.93 \text{ m}}{1 \times \tan 40°}$$

$$= 8.72 \text{ t/m}^2$$

This result is in accordance with the stresses already calculated as 10.70 and 12.875 t/m^2.

1.4 DESIGN OF WOODEN SUPPORTS

1.4.1 Design Principles

The design of wood support systems follows certain steps. First, the system is schematized and simple static models are drawn. Then, pressure evaluation is derived from many formulas, as previously demonstrated. The moment diagrams, maximum moments, maximum shear stresses, and the sections subjected to these moments and shears are calculated and dimensions determined. If these dimensions are found to be too large, modifications are made. Finally, verifications of allowable stresses are made for the dimensions and material. If the safety limits are not satisfied larger dimensions are chosen and calculations repeated until lower values of stresses allow safe usage.

1.4.2 Wooden Gallery Sets

The design of gallery sets consists of finding the proper size for caps, side posts, and auxilliaries such as wedges, laggings, and so forth. A

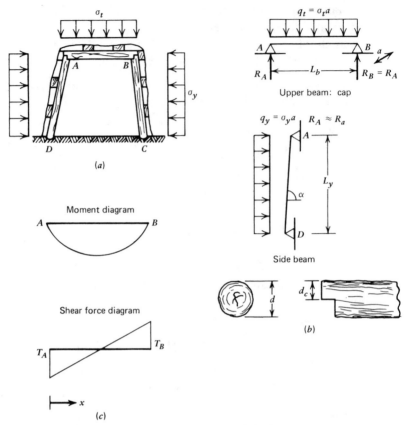

Figure 1.29 Design scheme of wooden gallery sets [2]: (a) supporting system; (b) static models; (c) simple beam diagrams.

typical wooden gallery set is shown in Fig. 1.29 [2, p. 398]. The stresses on caps and posts are shown with proper dimensions, and the moment and shear diagrams are incorporated. The wooden set works as a simple beam supported at both ends, loaded uniformly. Quantities and equations that apply to the figure are as follows:

$$M_x = 0.5 \, q_t L_b x - 0.5 \, q_t x^2$$

$$M_{\max} = 0.125 \, q_t L_b^2$$

$$x = \tfrac{1}{2} L_b$$

$$T_x = \frac{\partial M}{\partial x} = 0.5 \, q_t L_b - q_t x$$

Design of Wooden Caps.   The cap in wood supports is subjected to bending. The maximum bending moment and stress are given in the following formulas:

$$q_t = \sigma_t a \tag{1.36}$$

$$M_{max} = 0.125 \, q_t L_b^2 \tag{1.37}$$

$$\sigma_b = \frac{M_{max}}{W} \leqslant \sigma_{sf} \tag{1.38}$$

$$W = 0.098 \, d_b^3 \tag{1.39}$$

$$d_b \geqslant 1.084 \left(\frac{q_t}{\sigma_{sf}} L_b^2 \right)^{1/3} \tag{1.40}$$

where q_t = uniform load
  σ_t = uniform pressure
  a = distance between sets
M_{max} = maximum bending moment
  L_b = length of cap
  σ_b = bending stress
  σ_{sf} = allowable bending stress for wood
  d_b = diameter of cap

The load can be determined by

$$q_t = \alpha a \gamma L_b \tag{1.41}$$

$$d_b = 1.084 \, L_b \left(\frac{\alpha a \gamma}{\sigma_{sf}} \right)^{1/3} \tag{1.42}$$

For normal conditions we may take $\alpha = 0.5$, $\gamma = 0.0025$ kg/m^3, then

$$d_b = 0.117 \, L_b \left(\frac{a}{\sigma_{sf}} \right)^{1/3} \tag{1.43}$$

where d_b = diameter of cap, in centimeters
  L_b = length of cap, in centimeters
  a = distance between sets, in centimeters
  σ_{sf} = allowable bending stress, in kilograms per square centimeter, for second-class wood, 110 kg/cm^2

When designing the cap, if the distance between sets is 100 cm, and the allowable bending stress for second-class wood is 90 kg/cm^2, the diameter of the cap may be plotted against the length of the cap as shown in Fig. 1.30 [2, p. 400]. For light ($\alpha = 0.25$), medium ($\alpha = 0.5$) and bad roof conditions ($\alpha = 1$), three lines are shown. Since timber of a diameter larger than 25 cm is hard to get and difficult to handle, for bad conditions, and sets wider than 1.5 m, the distance between sets should be reduced (Fig. 1.30).

If we choose for the cap a definite diameter and find a proper length, we must verify the diameter against shear strength developed at the corners. At the corners, caps are cut to fit the post as shown in Fig. 1.29. We must consider this reduction in diameter, as well, for shear verification.

$$\tau_{max} = K \frac{T}{F} \leqslant \tau_{sf} \tag{1.44}$$

$$T = 0.5 \, q_t L_b = 0.5 \, \sigma_t L_b a \tag{1.45}$$

$$F = 0.785 \, d_b^2 \tag{1.46}$$

$$\tau_{max} = \frac{4}{3} \frac{0.5 \, \sigma_t L_b a}{0.785 \, d_b^2} \tag{1.47}$$

$$= 0.849 \frac{\sigma_t L_b a}{d_c^2} \frac{d_b}{d_c} \tag{1.48}$$

$$= 0.849 \frac{\sigma_t L_b a}{d_c^3} d_b \tag{1.49}$$

where τ_{max} = maximum shear stress, in kilograms per square centimeter

    K = factor, circular cross section, $\frac{4}{3}$

    T = maximum shear force, reactions at the corners, in kilograms

    F = cross-section area of the cap, in square centimeters

    d_b = diameter of cap, in centimeters

    d_c = diameter of cut on the cap at the corner (centimeters), d_c/d_b is the factor of corner fitting

    τ_{sf} = allowable shear stress, in kilograms per square centimeter

If the diameter found does not verify τ_{max}, it should be enlarged or, a (the distance between sets) should be diminished accordingly.

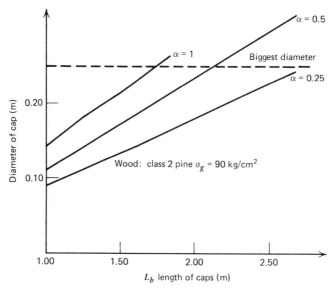

Figure 1.30 Diameter of caps plotted against the gallery width [2].

Design of Side Posts. The side posts of wooden supports are under side pressure and end reactions. Therefore in design normal compressive and bending stresses should be evaluated. In practice, the same diameters as are used as for caps. This diameter should be verified. The formulas involved are as follows:

$$\sigma_{sf} \geqslant \sigma_n \pm \sigma_b \tag{1.50}$$

$$\sigma_{sf} \overset{\geq}{=} -\frac{\omega R}{F} \pm 0.85 \frac{M_{max}}{W} \tag{1.51}$$

$$F = \frac{\pi}{4} d_y^2 = 0.785\, d_y^2 \tag{1.52}$$

$$M_{max} = 0.125\, q_y\, L_y^2 \tag{1.53}$$

$$W = 0.098\, d_y^3 \tag{1.54}$$

$$\lambda = \frac{4 L_k}{d_y} = \frac{4 L_y}{d_y} \tag{1.55}$$

$$\omega = f(\lambda)$$

$$R \cong 0.5\, q_t L_b \tag{1.56}$$

$$\sigma_{sf} \geqq -0.637\,\omega\,\frac{q_t L_b}{d_y^2} \pm 1.084\,\frac{q_y L_y^2}{d_y^3} \tag{1.57}$$

$$\sigma_{sf} \geqq -0.637\,\omega\,\frac{\sigma_t a L_b}{d_y^2} \pm 1.084\,\frac{\sigma_y a L_y^2}{d_y^3} \tag{1.58}$$

where σ_{sf} = allowable stress, in kilograms per square centimeter

σ_n = normal stress

σ_b = bending stress

ω = buckling factor (see Table 1.13), a function of slenderness

λ = number of slenderness

W = section modulus of post, in cubic centimeters

R = reaction load, in kilograms (although posts are slightly inclined they are taken as vertical)

q_t = uniform roof load, in kilograms per meter

σ_y = side pressures, in kilograms per square centimeter

L_b = length of cap, in centimeters

L_y = length of post, in centimeters

a = distance between sets, in centimeters

d_y = diameter of post, in centimeters

L_k = length for buckling = L_y

The buckling factor is obtained from Table 1.13, calculating λ from Eq. (1.55).

If the verification of Eq. (1.58) is made and found satisfactory, the d_y is determined. Otherwise, either a larger diameter or smaller a (the distance between sets) are chosen, and another trial is carried out.

Design of Wedges.   Wedges are designed in a manner similar to the design of caps. The spacing in ordinary conditions is quite sufficient. Under bad or changing conditions a new design should be made. Usually wedges cut longitidually from 12–18-cm props are quite sufficient. The design is done with the assumption that bending stress is under the safe limit (Fig. 1.31) [2, p. 404].

$$r = 1.142\,a\left(\frac{\sigma_y}{\sigma_{sf}}\right)^{1/2} \qquad \text{(wedge side by side)} \tag{1.59}$$

Table 1.13 Factors of Buckling ω[a]

λ	$\lambda+$										λ
	0	1	2	3	4	5	6	7	8	9	
0	1.00	1.01	1.01	1.02	1.03	1.03	1.04	1.05	1.06	1.06	0
10	1.07	1.08	1.09	1.09	1.10	1.11	1.12	1.13	1.14	1.15	10
20	1.15	1.16	1.17	1.18	1.19	1.20	1.21	1.23	1.23	1.24	20
30	1.25	1.26	1.27	1.29	1.29	1.30	1.32	1.33	1.34	1.35	30
40	1.36	1.38	1.39	1.40	1.42	1.43	1.44	1.46	1.47	1.49	40
50	1.50	1.52	1.53	1.55	1.56	1.58	1.60	1.61	1.63	1.65	50
60	1.67	1.69	1.70	1.72	1.74	1.76	1.79	1.81	1.83	1.85	60
70	1.87	1.90	1.92	1.95	1.97	2.00	2.03	2.05	2.08	2.11	70
80	2.14	2.17	2.21	2.24	2.27	2.31	2.34	2.38	2.42	2.46	80
90	2.50	2.54	2.58	2.63	2.68	2.73	2.78	2.83	2.88	2.94	90
100	3.00	3.07	3.14	3.21	3.28	3.35	3.43	3.50	3.57	3.65	100
110	3.73	3.81	3.89	3.97	4.05	4.13	4.21	4.29	4.38	4.46	110
120	4.55	4.64	4.73	4.82	4.91	5.00	5.09	51.9	5.28	5.38	120
130	5.48	5.57	5.67	5.77	5.88	5.98	6.08	6.19	6.29	6.40	130
140	6.51	6.62	6.73	6.84	6.95	7.07	7.18	7.30	7.41	7.53	140
150	7.65	7.77	7.90	8.02	8.14	8.27	8.39	8.52	8.65	8.78	150
160	8.91	9.04	9.18	9.31	9.45	9.58	9.72	9.86	10.00	10.15	160
170	10.29	10.43	10.58	10.73	10.88	11.03	11.18	11.33	11.48	11.64	170
180	11.80	11.95	12.11	12.27	12.44	12.60	12.76	12.93	13.09	13.26	180
190	13.43	13.61	13.78	13.95	14.12	14.30	14.48	14.66	14.84	15.03	190
200	15.20	15.38	15.57	15.76	15.95	16.14	16.33	16.52	16.71	16.91	200
210	17.11	17.31	17.51	17.71	17.92	18.12	18.33	18.53	18.74	18.95	210
220	19.17	19.38	19.60	19.81	20.03	20.25	20.47	20.69	20.92	21.14	220
230	21.37	21.60	21.83	22.06	22.30	22.53	22.77	23.01	23.25	23.49	230
240	23.73	23.98	24.22	24.47	24.72	24.97	25.22	25.48	25.73	25.99	240
250	26.25										250

[a]See reference 2.

$$r = 0.868\left(\frac{\sigma_y\, ca^2}{\sigma_{sf}}\right)^{1/3} \qquad \text{(wedge spaced)} \qquad (1.60)$$

$$h_k = 0.865\, a\left(\frac{\sigma_y}{\sigma_{sf}}\right)^{1/2} \qquad (1.61)$$

where r = thickness of half prop ($b = 2r$), in centimeters
h_k = thickness of a rectangular wedge, in centimeters
a = distance between gallery sets, in centimeters
c = distance between wedges, in centimeters

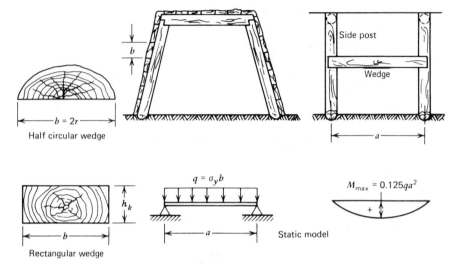

Figure 1.31 Designs for wedges on gallery set [2].

σ_y = side pressure, in kilograms per square centimeter (or σ_t roof pressure)

σ_{sf} = allowable bending stress, in kilograms per square centimeter

Numerical Application. Calculate the dimensions of a wooden gallery set under the following conditions:

Width of gallery = 1.75 m
Height = 2.00 m
Distance between sets = 0.75 m
Allowable bending stress for pine wood σ_{sf} = 110 kg/cm² (class 2 quality wood, see Table 1.8)
Allowable shear stress for pine wood τ_{sf} = 30 kg/cm³
Conditions of loading = medium (α = 0.5)

Let us first calculate pressure involved:

$$\text{Roof pressure} \quad \sigma_t = \alpha\gamma L_a$$

$$= 0.5 \times 2.5 \text{ t/m}^3 \times 1.75 \text{ m}$$

$$= 2.1875 \text{ t/m}^2 = 0.21875 \text{ kg/cm}^2$$

Side pressure $\sigma_y = K\sigma_t = 1 \times \sigma_t$

$$= 0.21875 \text{ kg/cm}^2$$

We can now calculate the diameter of the cap as follows:

$$d_b = 0.117 \, L_b \left(\frac{a}{\sigma_{sf}}\right)^{1/3}$$

$$= 0.117 \times 175 \left(\frac{75}{110}\right)^{1/3}$$

$$= 18 \text{ cm}$$

This diameter should be verified in respect to shear stress at the corners. Assuming that the cut for fitting is 12.5 cm, Eq. (1.49) may be used as follows:

$$\tau_{max} = \frac{0.849 \times 0.21875 \times 175 \times 75}{(12.5)^3} \quad (18)$$

$$= 22.46 \text{ kg/cm}^2$$

$$\leqslant 30 \text{ kg/cm}^2 \quad (\tau_{sf})$$

which is quite safe.

If we assume an 18-cm prop for side posts, we must then verify the buckling as follows:

$$\lambda = 4 \frac{L_k}{d_y} = 4 \, \frac{200}{18} = 44.44$$

If in Table 1.13 we take 40 from the first verticle column and 4 horizontally, $\omega = 1.42$; if we take 5 horizontally, $\omega = 1.43$. The total stress from bending and buckling is determined by Eq. (1.58) as follows:

$$\sigma = -0.637 \times 1.43 \times \frac{0.21875 \times 75 \times 175}{18^2}$$

$$\pm 1.084 \times \frac{0.21875 \times 75 \, (200)^2}{18^3}$$

$$= -8.08 \pm 121.98 \text{ kg/cm}^2$$

$$\sigma' = -8.08 - 121.98 = -130.06 \text{ kg/cm}^2$$

$$\sigma'' = -8.08 + 121.98 = +113.90 \text{ kg/cm}^2$$

Both of results are a little higher than the 110-kg/cm² allowable stress. In this case, a higher diameter (d = 20) may be chosen or the distance between the sets (a = 75 cm) may be reduced. Or, because experience has shown a leaning toward the safety factor, this size (d = 18) may be kept, since the difference between 130 and 114 kg/cm² is not too great. If excessive breakage occurs, the size may be increased.

The diameter should also be verified in respect to intrusion of the floor formation. The stress at the bottom of the posts is calculated as follows:

$$\sigma = \frac{\text{load}}{\text{area}} = \frac{0.5\, q_t L_b}{(\pi/4)\, d_y^2} = \frac{0.5\, \sigma_t a L_b}{(\pi/4)\, d_y^2}$$

$$= \frac{0.5 \times 0.21875 \times 75 \times 175}{0.785\,(18)^2}$$

$$= 5.64 \text{ kg/cm}^2$$

The least stable ground, shale, has a bearing capacity of 40 kg/cm². So the stress due to posts is quite low, and thus safe in respect to floor penetration.

The size of wedges, Eq. (1.59), assuming that they have been put at 40-cm intervals, is determined as follows:

$$r = 0.865 \left(\frac{0.21875 \times 40 \times (75)^2}{110} \right)^{1/3}$$

$$= 6.6 \text{ cm}$$

Thus we may split 12-cm posts in two parts longitudually (r = 6 cm) and use them as wedges.

The total consumption of wood per 1 m of gallery length can be calculated as follows:

1 cap	$(\pi/4)(0.18)^2\ 1.75$	0.044 m³
2 posts	$2(\pi/4)(0.18)^2\ 2.00$	0.101 m³
15 wedges	$15 \times \frac{1}{2} \times (\pi/4)(0.12)^2 0.75$	0.063 m³
	Total per set	0.208 m³

This table is for one set spaced at 0.75-m intervals; wood consumption per 1 m of gallery would be then calculated as follows:

$$\text{Consumption per 1 m gallery} = \frac{0.208}{0.75}$$

$$= 0.277 \text{ m}^3/\text{m}$$

1.4.3 Additions to Gallery Sets

Wide gallery sets usually require additions to diminish the size of caps and posts. Typical additions are shown in Fig. 1.32 [2, p. 409].

A cap with such additions works like an uniformly loaded beam with three supports. If we assume a medium loading condition

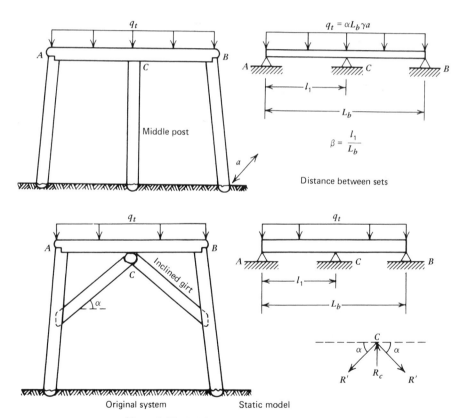

Figure 1.32 Additions to the gallery set [2].

($\alpha = 0.5$), the moments and reactions are given as follows:

$$M_c = -0.156\, a(3\beta^2 - 3\beta + 1)\, L_b^3 \qquad (1.62)$$

$$R_a = (0.157\, \beta^2 + 0.468\, \beta - 0.156)\,\frac{a}{\beta}\, L_b^2 \qquad (1.63)$$

$$R_b = (0.157\, \beta^2 - 0.785\, \beta + 0.468)\,\frac{a}{1-\beta}\, L_b^2 \qquad (1.64)$$

$$R_c = 1.25\, aL_b^2 - (R_a + R_b) \qquad (1.65)$$

$$\beta = \frac{l_1}{L_b}$$

where

$\qquad\qquad M_c$ = bending moment at middle post, in tonnes per meter

R_a, R_b, R_c = reaction forces at A, B, C in tonnes

$\qquad\qquad a$ = distance between sets, in meters

$\qquad\qquad L_b$ = length of the cap, in meters

$\qquad\qquad \beta$ = ratio where the central post is placed

The relations given by Eqs. 1.62–1.65 are shown graphically in Fig. 1.33 for convenience of design [2, p. 412].

As shown in Fig. 1.33, the least moment occurs at $\beta = 0.5$, the middle of the set, minimizing the size requirement for the timber. If girts are placed at an angle of 45°, the following numerical application can be made, if we assume the allowable bending stress to be 110 kg/cm² (1100 t/m²), the distance between the gallery sets to be $a = 1$ m, and gallery width to be $L_b = 3$ m.

$$M_c = 0.039\, aL_b^3$$

$$= 0.039 \times 1 \times (3)^3 = 1.053 \text{ t/m.}$$

Diameter of the cap d_b is determined as follows:

$$\sigma = \sigma_{sf} = 1100 = \frac{M_c}{W}$$

$$1100 = \frac{1.053}{0.098\, d_b^3} \text{ t/m}^2$$

$$d_b = \left(\frac{1.053}{0.098 \times 1100}\right)^{1/3}$$

$$= 0.21 \sim 20 \text{ cm}$$

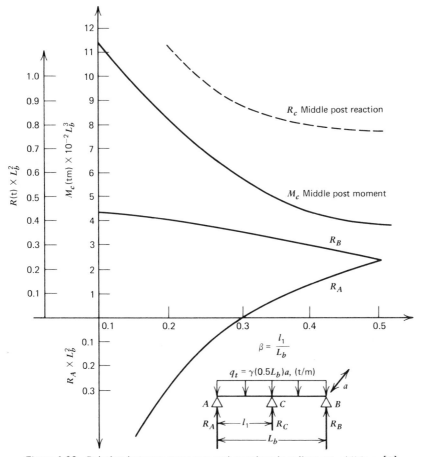

Figure 1.33 Relation between moments and reactions in gallery set additions [2].

If we do not use a central post, the size of the timber should be*

$$\sigma_{sf} = \frac{M_{max}}{W} = \frac{(q_t L_b^2)/8}{0.098 \, d^3} = \frac{0.5 \times 3.0 \times 2.5 \times 1 \times 3^2}{8 \times 0.098 \, d^3} = 1100 \text{ t/m}^2$$

$$d_b = (0.039)^{1/3} \cong 0.34 \text{ m}$$

This size is difficult to get and use in the mines. The necessity of additions to galleries wider than 2 m is self-explanatory.

*$q_t = \alpha L_b \gamma a = 0.5 \times 3 \text{ m} \times 2.5 \text{ t/m}^3 = 3.75 \text{ t/m}^2$.

The size of inclined girts required to work under conditions of compression and buckling can be calculated using reactions from Eq. (1.65) as follows:

$$R_c = 0.781 \, aL_b^2 = 0.781 \times 1 \times 3^2 = 7.029 \text{ t}$$

Inclined reactions R' are

$$R' = \frac{R_c}{2 \sin 45°} = \frac{7.029}{2 \times 0.707} = 4.97 \text{ t}$$

If the diameter is assumed to be 0.1 m, the compression and buckling stresses should be 85 kg/cm² (850 t/m²).

$$\sigma = \omega \frac{R}{\text{area}} = \omega \frac{4.97}{0.785 \, d_c^2} = 850 \text{ t/m}^2$$

The buckling number λ is calculated as follows:

$$\lambda = \frac{4l}{d_c} \cong \frac{4L_b}{2d_c \cos \alpha} = \frac{4 \times 3}{2 \times 0.1 \times 0.707} = 84.8 \cong 85$$

where (l is the length of the inclined girt.)

From Table 1.13 we find that ω is 2.31 for $\lambda = 85$ and

$$\sigma = 2.31 \frac{4.97}{0.785 \times (0.1)^2} = 1463 \text{ t/m}^2$$

which is greater than the allowable buckling stress of 850 t/m². Therefore the assumed diameter size of 0.1 m is too small. Taking side of $d_c = 0.125$ m, $\lambda \cong 68$, $\omega = 1.83$ we find

$$\sigma = 1.83 \frac{4.97}{0.785 \, (0.125)^2} = 742 \text{ t/m}^2$$

which is acceptable. The size of side posts should be the same as the caps.

1.4.4   Optimum Design

The sizes of caps, posts, and wedges are calculated as demonstrated in foregoing sections. For economy of design, size and spacing should be chosen to minimize expenditure of timber, that is, we should find the size (cap diameter) and spacing that necessitates the least volume of timber. The volume of caps in 1 m of distance is

$$V_b = \frac{\pi}{4} d_b^2 L_b \frac{100}{a} \tag{1.66}$$

$$d_b = 1.084 L_b \left(\frac{\alpha a \gamma}{\sigma_{sf}}\right)^{1/3} \qquad \text{see Eq. (1.42)}$$

$$a = \frac{d_b^3}{(1.084)^3 L_b^3} \frac{\sigma_{sf}}{\alpha \gamma} \tag{1.67}$$

$$V_b = \frac{100\,\pi}{4} (1.084)^3 L_b^4 \frac{\alpha \gamma}{\sigma_{sf}} \frac{1}{d_b}$$

$$V_b = \frac{A}{d_b} \tag{1.68}$$

$$A = \frac{100\,\pi}{4} \times (1.084)^3 L_b^4 \frac{\alpha \gamma}{\sigma_{sf}}$$

$$\cong \frac{100\,\alpha \gamma}{\sigma_{sf}} L_b^4 \tag{1.69}$$

where V_b = volume of the cap, in cubic centimeters
d_b = diameter of cap, in centimeters
L_b = length of cap, in centimeters
a = spacing of sets (distance between sets) in centimeters
α = loading factor (0.5 normal conditions)
γ = density of roof (taken as 0.0025 kg/cm^3)
σ_{sf} = allowable bending stress of wood, in kilograms per square
centimeter

Equation (1.68) shows that the volume of caps to be used decreases with the diameter.

The requirement for wedges, rectangular in section and put side by side, can be calculated as follows:

$$V_k = L_b h_k\ 100\ \text{cm}^3 \tag{1.70}$$

$$h_k = 0.865\, a \left(\frac{\alpha L_b \gamma}{\sigma_{sf}}\right)^{1/2} \qquad \text{see Eq. (1.61)}$$

$$a = \frac{d_b^3}{(1.084)^3 L_b^3} \frac{\sigma_{sf}}{\alpha \gamma} \qquad \text{see Eq. (1.67)}$$

$$V_k = 67.90 \; \frac{\sigma_{sf}}{\alpha\gamma L_b^2} \left(\frac{\alpha\gamma L_b}{\sigma_{sf}}\right)^{1/2} d_b^3 = Bd_b^3 \qquad (1.71)$$

$$B \cong 67.90 \; \frac{\sigma_{sf}}{\alpha\gamma L_b^2} \left(\frac{\alpha\gamma L_b}{\sigma_{sf}}\right)^{1/2} \qquad (1.72)$$

where V_k = volume of wood in wedges, in cubic centimeters
   h_k = thickness of wedges, in centimeters
   a = spacing of sets, in centimeters
   α = loading factor (0.5 normal conditions)
   γ = density of rock (0.0025 kg/cm^3)
   σ_t = pressure on the roof, in kilograms per square centimeter
   σ_{sf} = allowable bending stress in timber (class 2 wood), 110 kg/cm^2
   d_b = diameter of caps, in centimeters

Finally, the total timber consumption is determined as follows:

$$V = V_b + V_k = \frac{A}{d_b} + Bd_b^3 \qquad (1.73)$$

As one term decreases the other increases, so the sum comes to a minimum as shown in Fig. 1.34 [2, p. 418].

The minimum cap diameter can be found by differentiating

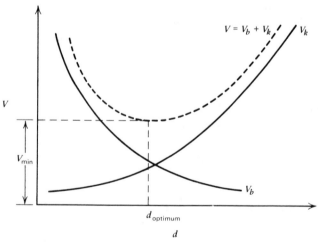

Figure 1.34   Timber consumption in terms of diameter of cap [2].

Eq. (1.73) as follows:

$$\frac{\partial V}{\partial d_b} = 0 \tag{1.74}$$

$$= Ad^{-2} + 3Bd_b^2 = 0 \tag{1.75}$$

$$d_b^4 = \frac{A}{3B}$$

$$d_b = \left(\frac{A}{3B}\right)^{1/4} \tag{1.76}$$

As a practical illustration, let us find the most economical dimensions for wooden gallery sets, 2 m wide at normal loading conditions ($\alpha = 0.5$), with allowable bending stress of 110 kg/cm^2, where the density of roof rock is 2.5 t/m^3.

First we should find A and B values as follows:

$$A \cong \frac{100 \times 0.5 \times 0.0025}{110} (200)^4 = 1818181$$

$$B = 67.90 \frac{110}{0.5 \times 0.0025 \times (200)^2}$$

$$\cdot \left(\frac{0.5 \times 0.0025 \times 200}{110}\right)^{1/2} = 7.12$$

$$d_b = \left(\frac{1818181}{3 \times 7.12}\right)^{1/4} = 17 \text{ cm}$$

$$a = \frac{(17)^3}{(1.084)^3 \times (200)^3} \left(\frac{110}{0.5 \times 0.0025}\right) = 42.4 \text{ cm}$$

$$h_k = 0.865 \times 42.4 \left(\frac{0.5 \times 200 \times 0.0025}{110}\right)^{1/2} \cong 2 \text{ cm.}$$

Thus the timber sets should be composed of caps and posts of 17 cm in diameter, spaced at 43 cm, wedged by 2-cm planks.

1.4.5 Design of Longwall Supports

The supports most frequently used in longwall faces are caps placed parallel to the face, in turn supported by three or four posts, as shown

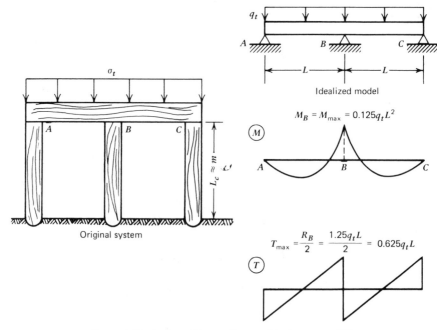

Figure 1.35 Design of longwall caps with three posts [2].

in Figs 1.35 and 1.36 [2, 421, 423]. It is assumed that the caps work like continuously loaded beams and that there is no sinking at the supports.

The design must consider the maximum moments and evaluate the ability of the cap diameter to take the bending moments within the allowable stress for wood. Then the shear stresses and stresses at floor are verified.

$$\sigma \leqslant \sigma_{sf} = 0.125 \frac{q_t L^2}{0.098 \, d^3} \qquad \text{(three posts)} \qquad (1.77)$$

$$d_b = 1.084 \left(\frac{\sigma_t a L^2}{\sigma_{sf}} \right)^{1/3} \qquad \text{(three posts)} \qquad (1.78)$$

$$d_b \cong \left(\frac{\sigma_t a L^2}{\sigma_{sf}} \right)^{1/3} \qquad \text{(four posts)} \qquad (1.79)$$

$$\tau = 1.061 \frac{\sigma_t L a}{d_b^2} \leqslant \tau_{sf} \qquad \text{(three posts)} \qquad (1.80)$$

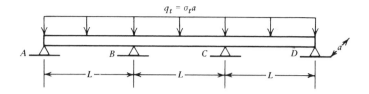

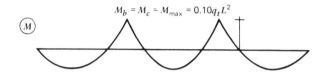

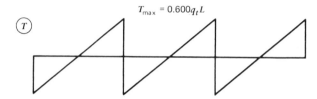

Figure 1.36   Design of longwall caps with four posts.

$$\tau = 1.019 \frac{\sigma_t La}{d_b^2} \leqslant \tau_{sf} \qquad \text{(four posts)} \qquad (1.81)$$

$$\sigma_c = \omega \frac{R_{max}}{F} = 1.59 \, \omega \, \frac{\sigma_t La}{d_b^2} \leqslant \sigma_{cs}$$

$$\text{(three posts)} \qquad (1.82)$$

$$\sigma_c = 1.40 \, \omega \, \frac{\sigma_t La}{d_b^2} \leqslant \sigma_{cs} \qquad \text{(four posts)} \qquad (1.83)$$

$$\lambda \cong \frac{4m}{d_b} \longrightarrow \omega = f(\lambda) \qquad \text{(table 1.13)}$$

$$\sigma_f = \frac{R_{max}}{F} = 1.59 \, \frac{\sigma_t La}{d_b^2} \qquad \text{(three post)} \qquad (1.84)$$

$$\sigma_f = 1.40 \, \frac{\sigma_t La}{d_b^2} \qquad \text{(four posts)} \qquad (1.85)$$

where σ_{sf} = allowable bending stress in wood (in tonnes per square meter)

σ_{cs} = allowable buckling stress in wood in tonnes per square meter

σ_f = stress in floor, in tonnes per square meter

q_t = uniform load, in tonnes per meter

L = distance between cap posts, in meters

d_b = diameter of caps and posts, in meters

a = distance between caps (width of cutting in shift), in meters

σ_t = roof pressure in the longwalls calculated by Eqs. 1.23, 1.30, or 1.34 (the largest figure may be taken for safest design)

ω = buckling factor

σ_c = compression stress parallel to fibers in post

As a practical example, let us calculate and evaluate the size of a cap at post four, caps being placed at 1.0-m intervals. The seam thickness is 1.5 m, and loading conditions are normal. The allowable bending stress is 1100 t/m^2, shear stress is 300, compression stress parallel to fibers is 850 t/m^2, and strength of rock is 1000 t/m^2.

$$h = \frac{1.8}{2.5 - 1.8} \times 1.5 = 3.86 \text{ m}$$

$$\sigma_t = h\gamma = 9.64 \text{ t/m}^2$$

$$d_b = \left(\frac{\sigma_t a L^2}{\sigma_{sf}} \right)^{1/3}$$

$$= \left(\frac{9.64 \times 1.0 \times 1.0^2}{1100} \right)^{1/3} = 0.2095 \text{ m}$$

$$\simeq 21 \text{ cm}$$

$$\tau = 1.019 \, \frac{\sigma_t a L}{d^2} = 1.019 \, \frac{9.64 \times 1 \times 1}{(0.21)^2}$$

$$= 222.7 < 300 \text{ t/m}^2$$

Thus the size is safe under the given shear conditions.

$$\lambda = \frac{4m}{d_b} = \frac{4 \times 1.50}{0.21} = 28.6$$

$$\omega = 1.24 \qquad \text{(Table 1.13)}$$

$$\sigma_c = 1.40\,\omega\,\frac{\sigma_t a L}{d_b^2}$$

$$= 1.40 \times 1.24\,\frac{9.64 \times 1 \times 1}{(0.21)^2}$$

$$= 379.48 < 850 \text{ t/m}^2$$

The calculation confirms that the posts are safe from buckling.

$$\sigma_f = 1.40\,\frac{\sigma_t L a}{d_b^2}$$

$$= 1.40\,\frac{9.64 \times 1.0 \times 1.0}{(0.21)^2}$$

$$= 306 < 1000 \text{ t/m}^2$$

The degree of penetration of the floor rock is also safe.

In practice the size of caps and posts are about 16 cm, less than the size calculated. This size works safely owing to the high safety factor (4–6) allowed for wood.

CHAPTER 2

Steel Gallery Supports

2.1 IMPORTANCE OF STEEL

The qualities of steel as a supporting material have caused it to replace wood in many mines, especially in coal mines where galleries are held for as long as 10 years for haulageways and air return. The basic characteristics of steel can be summarized as follows:

1. Steel is a very homogeneous material, metallurgically manufactured, free of natural defect, allowing lower safety factors to be used in designing.
2. Steel has a Young's modulus of elasticity (E = 2,000,000 kg/cm^2) much greater than any other structural materials giving it an advantage against deformations, buckling, and the like.
3. Steel can be manufactured in forms of alloys to meet high requirements set in designing.
4. Steel is the material least affected by atmospheric conditions such as temperature and humidity.
5. As a material, it can be reused by straightening. Completely deformed supports can be reclaimed as scrap.
6. On the other hand, it is an expensive material. Roadways supported by steel arches and the like are a big capital expense that smaller mines can not afford.

2.2   ENGINEERING CHARACTERISTICS OF STEEL

2.2.1   Chemical Structure

Chemically steel is an alloy of iron and carbon. There are some materials, like phosphorous (0.01–0.08%) and sulphur (0.01–0.06%), as impurities. Other materials, like manganese, silicon, nickel, chromium, and molybdenum, are incorporated in varying percentages to form special alloys to meet several conditions. For steel supports in the mines, ordinary steel of St. 37–52 is used and meets most structural requirements. Alloys are used for special conditions.

2.2.2   Mechanical Characteristics

A discussion of mechanical characteristics of steel must deal with stress-deformation, strength, modes of breaks, hardness, and design.

Stress-Deformation.   A typical stress-deformation curve is given in Fig. 2.1 [2, p. 433]. This is the ordinary curve where Young's modulus of elasticity is taken as $E = 2.1 \times 10^6$ kg/cm^2. The linear proportion continues to a point of 0.2% strain. After this point a "flow" takes place, with constant deformations, and failure occurs after these limits are reached.

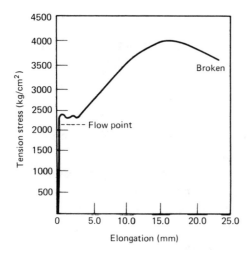

Figure 2.1   Stress–deformation curve of steel [2].

The tensile breaking strength of steel is given by the following empirical formula [23]:

$$\sigma = 0.00077 \left[38000 + C(700 + 2.94 \text{ Mn}) + 30 \text{ Mn} \right.$$

$$\left. + \frac{\text{Mn}}{200} (48 + 2.35 \text{ C}) + P(1000) + Si(340) \right] \qquad (2.1)$$

where σ = tensile breaking strength, in kilograms per square millimeter
  C = carbon in 0.01%
 Mn = manganese, 0.01%
  P = Phosphorous, 0.01%
  Si = Silicon, 0.01%

Carbon is the most important factor in tensile strength of steel elastic limits, and elongation at the breaking point. These properties are shown in Fig. 2.2 [23, 2].

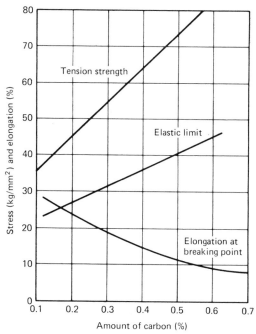

Figure 2.2 Effect of carbon in mechanical properties of steel [2, 23].

Table 2.1 Specifications of Steel According to DIN 21544[a]

Nomination	Elements (%) C	Mn	Flow Strength (kg/mm^2)	Average Tensile Strength (kg/mm^2)	Average Elongation (%)	Hardness (kg/cm^2)
St. 37	0.12	0.30	27	40	32	1.0
St. 42	0.12	0.40	29	46	30	1.0
St. 50	0.30	0.50	33	55	27	1.0
St. 52	0.18	1.10	U38	55	25	6.0
			V58	70	22	10.0
St. 54	0.35	0.55	34	60	25	1.3
St. 60	0.40	0.50	35	65	22	1.0
St. 70	0.40	0.75	45	75	15	0.9

[a]See references 2 and 23.

The mechanical strengths of various steels, according to DIN 21544 specifications, are rated in Table 2.1 [23, 2].

The physical properties of iron and steel used in structural work, according to the standards of American Society for Testing Materials, are given in Table 2.2 [24, 25 p. 43–42].

Mode of Failure.   Steel breaks in both ductile and brittle manners. In the ductile manner, breaking deformation is 100–200 times the flow

Table 2.2 Physical Properties of Iron and Steel[a]

Material	Weight (lb/ft^3)	Modulus of Elasticity[b] T and C	S (t)	Yield Point T and C	S (t)	Ultimate Strength T	C	S	Working Stress T	C	S
Iron: Gray, cast	450	15	6	–	–	25	100	25	4	16	4
Malleable, cast	475	22	8.8	–	–	45	110	45	8	20	8
Wrought	480	37	11	25	25	50	70	40	12	17	10
Steel: 0.1–0.2% carbon	490	30	12	35	20	60	90	48	18	18	12
0.3–0.4% "	490	30	12	40	24	80	45	64	20	20	16
0.7–0.8% "	490	30	12	60	36	125	70	85	30	30	21
Nickel, H T	490	30	12	85	55	112	95	80	25	25	20
Brass, rolled	520	15.5	6.2	25	17	73	30	47	18	15	11
Bronze, "	535	15	5.6	35	30	65	25	43	16	12	10
Aluminum, structural alloy	173	10	3.7	35	35	58	58	35	14.5	14.5	8.8

[a]See references 24 and 25. Values in thousands of lb per sq in. except as noted; T = tension; C = compression; S = shear; t = torsion.
[b]Millions of pounds per square inch.

deformation. The material reaches plastic deformation. This is usually seen in low carbon steels and is a desirable design characteristic.

Brittle failure is seen in high carbon steels where the deformation is rather small and the breaking surfaces are rough. There is no definite flow point.

Hardness. Hardness is a relative property, measured as the reaction toward intrusion. According to material science, the Brinell hardness is defined as the area of intrusion (square millimeters) by a spherical ball under a known force. According to the empirical formula [23]

$$\sigma_k = 0.34 \text{ HB} \qquad (2.2)$$

where σ_k = tensile breaking strength, in kilograms per square millimeter

HB = Brinell number

2.2.3 Characteristics of Support Elements

The characteristics to be considered in support elements are profile (cross-section area), strength moments, Rankin ratio, and allowable stresses.

Profile. The strength of a beam is proportional to the profile (cross-section area). The weight and value of the beam are also proportional to the profile. The use of heavy profiles has limitations in the mines which necessitates the use of light to medium-heavy materials. The section area and weight of I-beams of DIN 21541 are given in Table 2.3 [23, 2, p. 438], and American I-beams are given in Table 2.4 [24, 25]. The Toussiant-Heinzmann (T–H) cross sections are tabulated in Fig. 2.3 and Table 2.5, [26, 2, p. 449].

Moments of Inertia and Section Modulus. Tables 2.4 and 2.5 also give values for inertia I (in centimeters to the fourth power) and section modulus (in cubic centimeters). These values differ according to the x–x and y–y axes as shown in Fig. 2.3. and are used in designing.

Rankin Ratio. This is the ratio of compressive strength to buckling strength in a beam 2 m long. The ratio is always greater than 1, but it

Table 2.3 Characteristics of I-beams of DIN 21541[a]

Symbol	h	b	t_1	t_2	r_1	r_2	Incl.	Section F(cm²)	Unit Weight (kg/m)	x–x			y–y		
										I_x (cm⁴)	W_x cm³	i_x (cm)	I_y (cm⁴)	W_y (cm³)	i_y (cm)
GI 70	70	68	7	9.5	10	3	30	16.2	13.0	122	34.7	2.74	36.0	10.6	1.49
GI 90	90	76	8	11.5	12	4	30	22.5	17.7	281	62.5	3.53	62.6	16.5	1.67
GI 100	100	80	9	12.5	13	5	30	26.4	20.7	403	80.7	3.91	80.5	20.1	1.75
GI 110	110	84	10	14	14	6	33	31.1	24.5	570	103	4.28	103	24.5	1.82
GI 120	120	92	11	15.5	15	7	33	37.6	29.5	816	136	4.66	150	32.6	2.00
GI 130	130	100	12	17	16	8	33	44.6	35.0	1,130	175	5.05	211	42.3	2.18
GI 140	140	110	12	19	17	8	33	53.0	41.6	1,586	227	5.47	315	57.3	2.44

[a]See references 2 and 23.

Table 2.4 American Standard I-beams[a]

Nominal Size (in.)	Weight per Foot (lb)	Area of Section (in.²)	Depth of Section (in.)	Width of Flange (in.)	Web Thickness (in.)	Axis 1-1			Axis 2-2		
						I in⁴	S in³	r in	I in⁴	S in³	r in
$24 \times 7\frac{7}{8}$	120.0	35.13	24.00	8.048	0.798	3 010.8	250.9	9.26	84.9	21.1	1.56
	115.0	33.67	24.00	7.987	0.737	2 940.5	245.0	9.35	82.8	20.7	1.57
	110.0	32.18	24.00	7.925	0.675	2 869.1	239.1	9.44	80.6	20.3	1.58
	105.9	30.98	24.00	7.875	0.625	2 811.5	234.3	9.53	78.9	20.0	1.60
24×7	100.0	29.25	24.00	7.247	0.747	2 371.8	197.6	9.05	48.4	13.4	1.29
	95.0	27.79	24.00	7.186	0.686	2 301.5	191.8	9.08	47.0	13.0	1.30
	90.0	26.30	24.00	7.124	0.621	2 230.1	185.8	9.21	45.5	12.8	1.32
	85.0	24.84	24.00	7.063	0.563	2 159.8	180.0	9.33	44.2	12.5	1.33
	79.9	23.33	24.00	7.000	0.500	2 087.2	173.9	9.46	42.9	12.2	1.36
20×7	100.0	29.20	20.00	7.273	0.873	1 648.3	164.8	7.51	52.4	14.4	1.34
	95.0	27.74	20.00	7.200	0.800	1 599.7	160.0	7.59	50.5	14.0	1.35
	90.0	26.26	20.00	7.126	0.726	1 550.3	155.0	7.68	48.7	13.7	1.36
	85.0	24.80	20.00	7.053	0.653	1 501.7	150.2	7.78	47.0	13.3	1.38
	81.4	23.74	20.00	7.000	0.600	1 466.3	146.6	7.86	45.8	13.1	1.39
$20 \times 6\frac{1}{4}$	75.0	21.90	20.00	6.391	0.641	1 263.5	126.3	7.60	30.1	9.4	1.17
	70.0	20.42	20.00	6.317	0.567	1 214.2	121.4	7.71	28.9	9.2	1.19
	65.4	19.08	20.00	6.250	0.500	1 169.5	116.9	7.83	27.9	8.9	1.21
18×6	70.0	20.46	18.00	6.251	0.711	917.5	101.9	6.70	24.5	7.8	1.09
	65.0	18.98	18.00	6.169	0.629	877.7	97.5	6.80	23.4	7.6	1.11
	60.0	17.50	18.00	6.087	0.547	837.8	93.1	6.92	22.3	7.3	1.13
	54.7	15.94	18.00	6.000	0.460	795.5	88.4	7.07	21.2	7.1	1.15
15×6	75.0	21.85	15.00	6.278	0.868	687.2	91.6	5.61	30.6	9.8	1.18
	70.0	20.38	15.00	6.180	0.770	659.6	87.9	5.69	28.8	9.3	1.19
	65.0	18.91	15.00	6.082	0.672	632.1	84.3	5.78	27.2	8.9	1.20
	60.8	17.68	15.00	6.000	0.590	609.0	81.2	5.87	26.0	8.7	1.21
$15 \times 5\frac{1}{2}$	55.0	16.06	15.00	5.738	0.648	508.7	67.8	5.63	17.0	5.9	1.03
	50.0	14.59	15.00	5.640	0.550	481.1	64.2	5.74	16.0	5.7	1.05
	45.0	13.12	15.00	5.542	0.452	453.6	60.5	5.88	15.0	5.4	1.07
	42.9	12.49	15.00	5.500	0.410	441.8	58.9	5.95	14.6	5.3	1.08
$12 \times 5\frac{1}{4}$	55.0	16.04	12.00	5.600	0.810	319.3	53.2	4.46	17.3	6.2	1.04
	50.0	14.57	12.00	5.477	0.687	301.6	50.3	4.55	16.0	5.8	1.05
	45.0	13.10	12.00	5.355	0.565	284.1	47.3	4.66	14.8	5.5	1.06
	40.8	11.84	12.00	5.250	0.460	268.9	44.8	4.77	13.8	5.3	1.08
12×5	35.0	10.20	12.00	5.078	0.428	227.0	37.8	4.72	10.0	3.9	0.99
	31.8	9.26	12.00	5.000	0.350	215.8	36.0	4.83	9.5	3.8	1.01
$10 \times 4\frac{3}{4}$	40.0	11.69	10.00	5.091	0.741	158.0	31.6	3.68	9.4	3.7	0.90
	35.0	10.22	10.00	4.944	0.594	145.8	29.2	3.78	8.5	3.4	0.91
	30.0	8.75	10.0	4.797	0.447	133.5	26.7	3.91	7.6	3.2	0.93
	25.4	7.38	10.00	4.660	0.310	122.1	24.4	4.07	6.9	3.0	0.97
8×4	25.5	7.43	8.00	4.262	0.532	68.1	17.0	3.03	4.7	2.2	0.80
	23.0	6.71	8.00	4.171	0.441	64.2	16.0	3.09	4.4	2.1	0.81
	20.5	5.97	8.00	4.079	0.349	60.2	15.1	3.18	4.0	2.0	0.82
	18.4	5.34	8.00	4.000	0.270	56.9	14.2	3.26	3.8	1.9	0.84
$7 \times 3\frac{3}{4}$	20.0	5.83	7.00	3.860	0.450	41.9	12.0	2.68	3.1	1.6	0.74
	17.5	5.09	7.00	3.755	0.345	38.9	11.1	2.77	2.9	1.6	0.76
	15.3	4.43	7.00	3.660	0.250	36.2	10.4	2.86	2.7	1.5	0.78

Table 2.4 (*Continued*)

Nominal Size (in.)	Weight per Foot (lb)	Area of Section (in.²)	Depth of Section (in.)	Width of Flange (in.)	Web Thickness (in.)	Axis 1–1 I in⁴	Axis 1–1 S in³	Axis 1–1 r in	Axis 2–2 I in⁴	Axis 2–2 S in³	Axis 2–2 r in
$6 \times 3\frac{3}{8}$	17.25	5.02	6.00	3.565	0.465	26.0	8.7	2.28	2.3	1.3	0.68
	14.75	4.29	6.00	3.443	0.343	23.8	7.9	2.36	2.1	1.2	0.69
	12.5	3.61	6.00	3.330	0.230	21.8	7.3	2.46	1.8	1.1	0.72
5×3	14.75	4.29	5.00	3.284	0.494	15.0	6.0	1.87	1.7	1.0	0.63
	12.25	3.56	5.00	3.137	0.347	13.5	5.4	1.95	1.4	0.91	0.63
	10.0	2.87	5.00	3.000	0.210	12.1	4.8	2.05	1.2	0.82	0.65
$4 \times 2\frac{3}{4}$	10.5	3.05	4.00	2.870	0.400	7.1	3.5	1.52	1.0	0.70	0.57
	9.5	2.76	4.00	2.796	0.326	6.7	3.3	1.56	0.91	0.65	0.58
	8.5	2.46	4.00	2.723	0.253	6.3	3.2	1.60	0.83	0.61	0.58
	7.7	2.21	4.00	2.660	0.190	6.0	3.0	1.64	0.77	0.58	0.59
$3 \times 2\frac{3}{8}$	7.5	2.17	3.00	2.509	0.349	2.9	1.9	1.15	0.59	0.47	0.52
	6.5	1.88	3.00	2.411	0.251	2.7	1.8	1.19	0.51	0.43	0.52
	5.7	1.64	3.00	2.330	0.170	2.5	1.7	1.23	0.46	0.40	0.53

[a]See references 24 and 25.

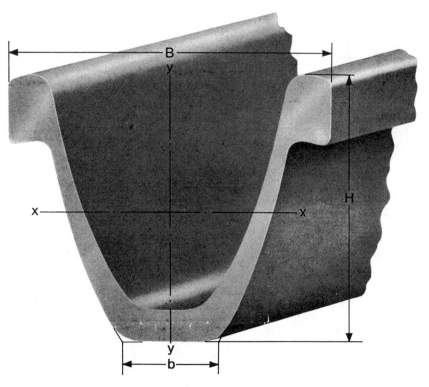

Figure 2.3 Perspective view of Toussaint–Heinzmann profiles [2, 26].

Table 2.5 Characteristics of Toussaint-Heinzmann Profiles[a]

Weight (kg/m)	13	16	21	25	29	36	44
Type (tip)	48	48	58	58	58	58	58
Height H(mm)	85	89	108	118	124	138	148
Width B(mm)	98	98	124	135	151	171	172
Area F(cm^2)	16	20	27	32	37	46	56
Weight G(kg/m)	13	16	21	25	129	36	44
Moment of inertia I_x(cm^4)	137	176	341	484	616	972	1265
Section modulus W_x(cm^3)	32	40	61	80	94	137	171

[a] See references 2 and 26.

is advantageous in designing to be close to 1. The Rankin ratio and section moduli W_x, W_y are seen in some beams, such as rail, Clement, and Toussaint–Heinzmann profiles, shown in Fig. 2.4 and Table 2.6.

Allowable Stress. Normal steel (St. 37) has an allowable stress of 1400 kg/cm^2 and a flow stress of 2400 kg/cm^2. The safety factor is

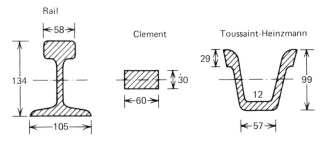

Figure 2.4 Dimensions of some profiles of beams [2].

Table 2.6 Characteristics of Some Profiles[a]

Characteristics	Rail	Clement	Toussaint-Heinzmann
Unit weight (kg/m)	33.5	14	21
W_x (cm^3)	155	7	58
W_y (cm^3)	50	14	63
Rankin ratio	1.5	5.3	1.3

[a] See reference 2.

2400/1400 = 1.71. For a higher quality steel (St. 52) the flow limit is 3600, 1.5 times that of St. 37. If such a steel is used in designing problems the allowable stress is

$$\sigma_{sf} = 1.5 \times 1400 = 2100 \text{ kg/cm}^2$$

which may be more economical in many supports.

2.3 DESIGN OF RIGID ARCHES

2.3.1 Description of Rigid Arches

Typical rigid steel arches of 10-m² and 18-m² cross-section area are shown in Fig. 2.5 with dimensions and connection details [2, p. 445–447]. They are half elliptical with the largest dimension on the floor. There is a minimum of 75 cm from the side wall to the top of the mine car to allow a safe space for a man to stand during passage of cars.

2.3.2 Stress Evaluation

Many rigid arches can be simplified in a half-circular shape above a vertical distance. The connecting parts are assumed to be very "rigid" and shown as "continuous," as in Fig. 2.5. The static analysis is given in Fig. 2.6 [27, 2 p. 470] :

$$A_y = B_y = \frac{(0.785 \ h' + 0.666 \ r) \ q_t r^3}{0.666 \ h'^3 + \pi r h'^2 + 4h'r^2 + 1.57 \ r^3} \tag{2.3}$$

$$M = 0.5 \ q_t r^2 \sin^2 \alpha - A_y (h' + r \sin \alpha) \qquad \text{for } 0 \leqslant \alpha < \pi \tag{2.4}$$

$$M = -A_y x \qquad \text{for } 0 \leqslant x \leqslant h' \tag{2.5}$$

$$N = -q_t r \cos^2 \alpha - A_y \sin \alpha \tag{2.6}$$

where $A_y = B_y$ side reactions, in tonnes

$\qquad h' =$ vertical distance of the arch, in meters

$\qquad r =$ radius of the arch, in meters

$\qquad \alpha =$ angle from horizontal, in degrees (see Fig. 2.6)

$\qquad q_t =$ uniform roof load, in tonnes per meter

$\qquad M =$ moment, in tonnes · meter

$\qquad N =$ normal force to the profile, in tonnes

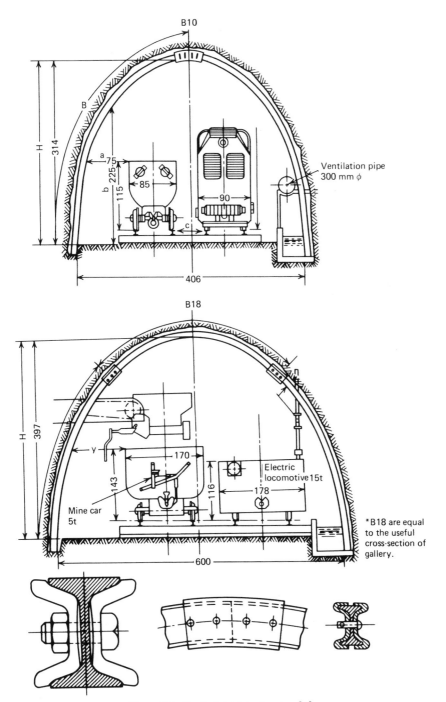

Figure 2.5 Typical rigid steel arches [2].

71

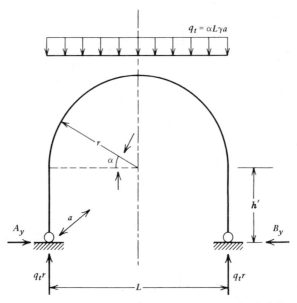

Figure 2.6 Static model of an idealized rigid steel arch [2].

To design rigid arches the maximum moment should be known. If we differentiate Eq. (2.4) with respect to α, and equate to zero, we have:

$$\frac{\partial M}{\partial \alpha} = \cos \alpha (q_t r^2 \sin \alpha - A_y r) = 0 \tag{2.7}$$

$$\cos \alpha = 0, \qquad \alpha = \frac{\pi}{2} \tag{2.8}$$

$$q_t r^2 \sin \alpha - A_y r = 0 \tag{2.9}$$

$$\sin \alpha = \frac{A_y}{q_t r}, \qquad \alpha = \sin^{-1} \frac{A_y}{q_t r} \tag{2.10}$$

The values of $M_{\max}$ and N are for values of α of Eqs. (2.8) and (2.10) as follows:

$$M_{\max} = 0.5 \, q_t r^2 - A_y (h' + r) \tag{2.11}$$

$$M_{\max} = -A_y \left(h' + 0.5 \frac{A_y}{q_t} \right) \tag{2.12}$$

$$N = -A_y \qquad (2.13)$$

$$N_1 = -q_t r \qquad (2.14)$$

The values of Eqs. (2.11) and (2.13) are much smaller than the values of Eqs. (2.12) and (2.14), respectively.

2.3.3 Design of Arch Profile

The values of Eqs. 2.12 and 2.14 should be used in calculating the cross section of the arch. The stress should be determined as follows:

$$|\sigma| = \frac{\text{normal load}}{\text{profile area}} + \frac{\text{maximum moment}}{\text{section modulus}}$$

$$|\sigma| = \frac{q_t r}{F} + \frac{A_y(h' + 0.5 \, A_y/q_t)}{W} \leqslant \sigma_{sf} \qquad (2.15)$$

where $|\sigma|$ = absolute stress value, in tonnes per square meter

F = section area of the profile, in square meters

W = section modulus of the profile, in cubic meters

σ_{sf} = allowable stress in steel for mine supports, 1400 kg/cm^2 or 14000 t/m^2

In Eq. (2.15) the cross section and the section modulus are two unknowns, so trial and error should be used for a proper design. However, in DIN specifications

$$F = 0.149 \, W + 9.780 \qquad (2.16)$$

and

$$|\sigma| = \frac{q_t r}{0.149 \, W + 9.780} + \frac{A_y(h' + 0.5 \, A_y/q_t)}{W} \leqslant \sigma_{sf} \qquad (2.17)$$

Equation (2.17) is second degree with respect to W, and the positive root of the equation should be taken. After W is determined the nearest profile is obtained from Table 2.3.

A more elaborate and precise way to evaluate stresses and the design of a proper beam, introduced by Proctor and White [28], is not included here as it is very complicated and usually applied to large tunnel supports. A numerical example is given by Peng [29, p. 409].

2.3.4 Numerical Application

Let us find the appropriate DIN profile for a rigid arch of a gallery 8 m² in section spaced at 1-m intervals, under the normal stress conditions. ($\alpha = 0.5$, $\gamma = 2.5$ t/m³). The data can be summarized as follows (Fig. 2.6):

L = span of gallery = 3.65 m

r = 1.675 m

h' = 1.20 m

a = 1.0 m (spacing of arches)

α = 0.5 (normal stress condition)

$q_t = \alpha L \gamma a$ as given in Eq. (1.17)

     = 0.5 × 3.65 m × 2.5 t/m³ × 1.0 m

     = 4.562 t/m

$$A_y = \frac{(0.785 \times 1.2 + 0.666 \times 1.675) \times 4.562(1.675)^3}{0.666(1.20)^3 + \pi(1.675)(1.2)^2 + 4 \times 1.2(1.675)^2 + 1.57(1.675)^3}$$

     = 1.491 t

$M = 0.5 \times 4.562(1.675)^2 \sin^2 \alpha - 1.491(1.2 + 1.675 \sin \alpha)$

$N = -4.562 \times 1.675 \cos^2 \alpha - 1.491 \sin \alpha$

To show the maximum values, M and N values are plotted in the polar coordinates of angle α and are shown in Table 2.7 [2, p. 472].

Table 2.7   Moments and Normal Load on a Rigid Steel Arch[a]

Angle Degrees	0	15°	30°	45°	60°	75°	90°
Moment (t · m)	-1.79	-2.0	-1.44	-0.36	0.38	1.77	2.11
Normal load (t)	7.64	-7.52	-6.48	-4.87	3.20	-1.95	-1.49

[a]See reference 2.

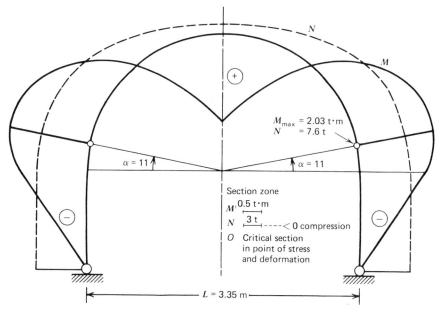

Figure 2.7 Moments and normal loads on a rigid steel arch plotted in polar coordinates [2].

The maximum values are as follows:

$$\alpha = \sin^{-1} \frac{A_y}{q_t r} = \sin^{-1} \frac{1.491}{4.562 \times 1.675} = \sin^{-1} 0.1951$$

$$\alpha = 11.25°$$

$$M_{\max} = -A_y \left(h' + 0.5 \frac{A_y}{q_t} \right)$$

$$= -1.491 \left(1.2 + 0.5 \frac{1.491}{4.562} \right)$$

$$= -2.03 \text{ t} \cdot \text{m}$$

$$N_1 = -q_t r = -4.562 \times 1.675$$

$$= -7.641 \text{ t}$$

These values are plotted on the polar coordinate in Fig. 2.7 [2, p. 473].
The proper I-beam according to Eq. (2.17) is calculated as follows:

$$\frac{4.652 \times 1.675}{0.149\,W + 9.78} + \frac{1.491[1.2 + 0.5(1.491/4.562)]}{W} = 14000$$

$$7.6414\,W + 2.0329(0.149\,W + 9.78) = 14000\,W(0.149\,W + 9.78)$$

$$2086\,W^2 + 136912.3257\,W - 19.8818 = 0$$

$$W_1 = 0.00014521\ \text{m}^3$$

$$= 145.21\ \text{cm}^3$$

From Table 2.3 we find that the GI 130 profile is suitable for the system under consideration.

2.4   DESIGN OF ARTICULATED (MOLL) ARCHES

2.4.1   Description of Articulated Arches

Several articulated arches used in the mines are shown in Fig. 2.8, and the forms of articulations of Moll arches are illustrated in Fig. 2.9 [2, p. 457].

The most popular of articulated arches, the "moll arches," are constructed from three long pieces of wooden caps with steel arched sections resting on these caps. The plan view of such an arched gallery is shown in Fig. 2.8a and the longitudinal cross section at Fig. 2.8b. In the figure, the wooden caps are designated by 1 and 2, placed on top and on sides of the gallery respectively, and the steel arches are designated by 3. The side caps are supported by either steel posts (Fig. 2.8c, designated by 4), wooden posts (Fig. 2.8d, designated by 5), or wooden chocks (Fig. 2.8e). In some cases "fillings" are used (Fig. 2.8f, g) to support the wooden caps; these are termed "sinking arches," as the filling crumbles and the arches lose height. To reduce timber consumption, the roof cap can be replaced by several steel articulations as shown in Fig. 2.9.

2.4.2   Design of a Moll Arch with Two Articulations

The cross section of an articulated moll arch supported by wooden chocks and the statics of the cross section are illustrated in Fig. 2.10 [2, p. 475].

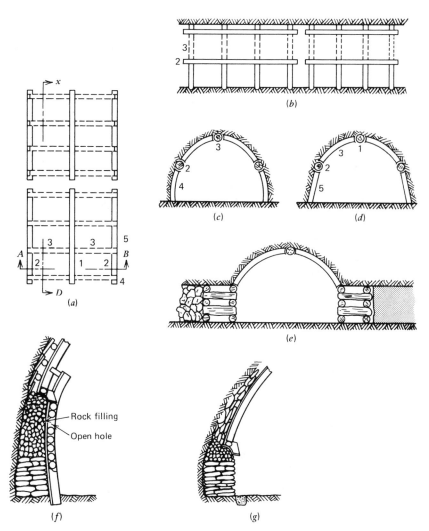

Figure 2.8 Forms of articulated arches [2].

Analysis of the two-articulation moll arch is quite similar to that rigid arch in Fig. 2.6, only the vertical portion is reduced to zero. So, Eqs. (2.3), (2.4), and (2.5) are modified by $h' = 0$, thus

$$A_y = B_y = \frac{0.666\, q_t r^4}{1.57\, r^3} = 0.424\, q_t r \qquad (2.18)$$

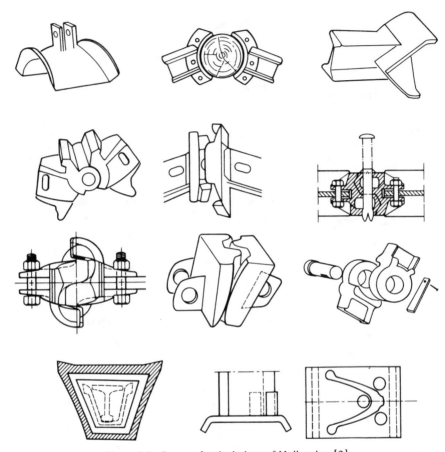

Figure 2.9 Forms of articulations of Moll arches [2].

$$M = 0.5\, q_t r^2 \sin^2 \alpha - 0.424\, q_t r(r \sin \alpha)$$
$$= q_t r^2 \sin \alpha(0.5 \sin \alpha - 0.424) \tag{2.19}$$
$$N = -q_t r \cos^2 \alpha - 0.424\, q_t r \sin \alpha$$
$$= -q_t r(\cos^2 \alpha + 0.424 \sin \alpha) \tag{2.20}$$

$$\frac{\partial M}{\partial \alpha} = \cos \alpha\, (\sin \alpha - 0.424) = 0$$

$$\cos \alpha = 0 \longrightarrow \alpha = \pi/2$$

$$\sin \alpha = 0.424, \quad \alpha = 25°$$

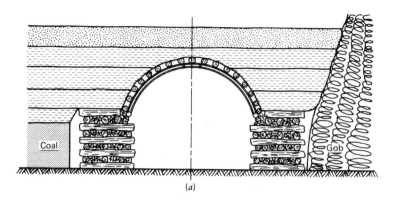

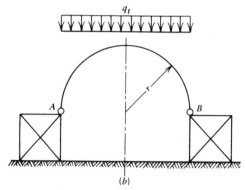

Figure 2.10 Moll arches with two articulations [2]: (*a*) typical support system; (*b*) idealized static model.

$$M_{\max} = 0.076 \, q_t r^2 \qquad \text{for } \alpha = \pi/2 \qquad (2.21)$$

$$M_{\max} = -0.09 \, q_t r^2 \qquad \text{for } \alpha = 25° \qquad (2.22)$$

The $\alpha = 25°$ values are greater in absolute, therefore the moments and normal forces are plotted against angle α in the polar coordinate, given in Fig. 2.11; these values are taken for design purposes. Then:

$$N_{25°} \cong -q_t r \qquad (2.23)$$

$$|\sigma| = \frac{q_t r}{F} + \frac{0.09 \, q_t r^2}{W} \qquad (2.24)$$

where F and W are the profile area and the section modulus of the profile section.

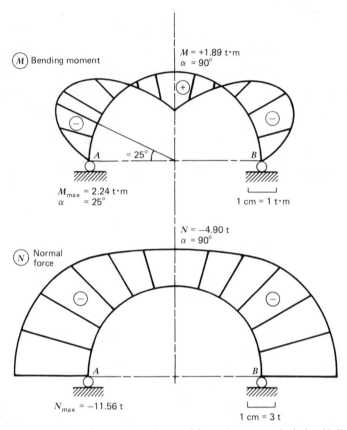

Figure 2.11 Distribution of moments and normal forces in a two-articulation Moll arch [2].

As a numerical example, let us calculate the size of moll arches 4.30 m wide spaced at 1-m intervals.

$$r = \frac{L}{2} = \frac{4.30}{2} = 2.15 \text{ m}$$

$$a = 1.0 \text{ m}$$

$$q_t = \alpha L \gamma a = 0.5 \times 4.30 \text{ m} \times 2.5 \text{ t/m}^3 \times 1.0 \text{ m}$$

$$= 5.375 \text{ t/m}$$

$$M_{max} = -0.09 \times 5.375 (2.15)^2 = -2.24 \text{ t} \cdot \text{m} = -224000 \text{ kg/cm}$$

$$N = -5.375 \times 2.15 = -11.56 \text{ t} = -11560 \text{ kg}$$

The easiest way is to take several profiles and verify the stresses allowed: Let us take DIN profile GI 110. Such a profile has a cross-section area of 31.1 cm² and section modulus of 103 cm³. Then

$$|\sigma| = \frac{11560}{31.1} + \frac{224000}{103}$$

$$= 2546 > \sigma_{sf} = 1400 \text{ kg/cm}^2$$

Under these conditions there are the following three possibilities:

Reduce the distance between supports (smaller a).

Use a larger profile (larger F and W).

Use a higher quality of steel (such as St. 52).

In our sample, if we take a profile GI = 140, $F = 53.0$ cm², $W = 227$ cm³ (Table 2.3), then

$$\sigma = \frac{11560}{53.0} + \frac{224000}{227.0} = 1204 < 1400 \text{ kg/cm}^2$$

which is quite suitable.

2.4.3   Design of a Moll Arch with Three Articulations

Figure 2.12 shows the statics of an arch with three articulations, and maximum values of moments and normal forces are given as follows:

$$M_{max} = 0.125 \, q_t r^2 \longrightarrow x = 0.134 \, r \qquad (2.25)$$

$$N_x = -q_t r \qquad\qquad x = 0.134 \, r \qquad (2.26)$$

$$|\sigma| = \frac{q_t r}{F} + \frac{0.125 \, q_t r^2}{W} \leqslant \sigma_{sf} \qquad (2.27)$$

Using the same numerical values as above with a GI = 140 profile, we have the following:

$$|\sigma| = \frac{53.75 \times 215}{53.0} + \frac{0.125 \times 53.75(215)^2}{227.0}$$

$$= 1590 > \sigma_{sf} = 1400 \text{ kg/cm}^2$$

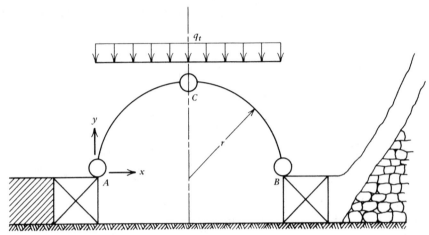

Figure 2.12   Three-articulation Moll system [2].

It can be seen that with three articulations higher values of moments should be met, and the profile used for two articulations is not safe. Either a larger profile must be used or the distance between sets must be reduced.

2.5   DESIGN OF YIELDING ARCHES

2.5.1   Description of Yielding Arches

Yielding arches are composed of three sections. The top section slides between two side elements. Every 15 days or so, the tightening elements are loosened and the arches slide, converging and thus relieving the stresses on them, eliminating deformations. This is shown schematically in Fig. 2.13a, b, c [2, p. 445].

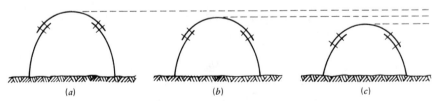

Figure 2.13   Working principle of yielding arches [2].

The first yielding arches were designed by Toussaint and Heinz-mann with U-shaped of profiles and are shown in Fig. 2.14 [26, 2, p. 448]. After the patent termination, other forms of yielding arches such as "Glocken" and "Kunstler" have been marketed with V-section and flat iron shapes put in the form of a U-section, as shown in Figs. 2.15 and 2.16 respectively [26, 2, p. 454–453].

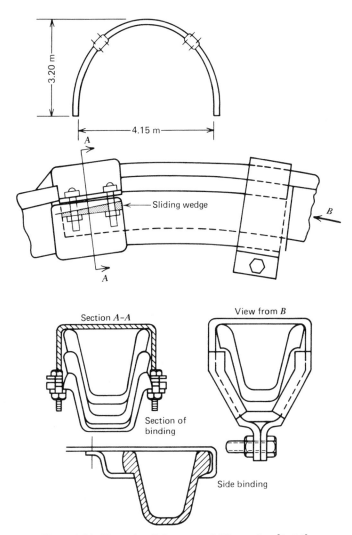

Figure 2.14 Toussaint–Heinzmann yielding arches [2, 26].

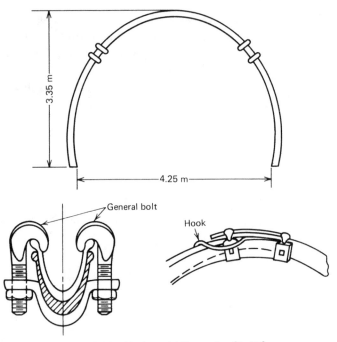

Figure 2.15   Glocken yielding arches [2, 26].

Yielding arches, used for the most part in the main and tailgates of the longwall panels, are not very large in cross section. Typical Toussaint–Heinzmann arches used in the French collieries are shown in Fig. 2.17, and their dimensions are summarized in Table 2.8 [30, 2, p. 451].

Table 2.8   Dimensions of French Toussiant-Heinzmann Arches[a]

Arch Number	Gallery Section (m²)	Arch Elements						
		L	H	P	d	R	C	r
P 250	5.3	2.78	2.48	2.4	0.76	2.0	2.6	1.2
P 300	7.86	2.81	3.34	2.6	0.85	2.2	3.1	1.5
P 370	9.45	3.84	2.99	2.9	0.78	2.4	3.1	1.7
P 420	11.20	4.34	3.14	3.1	0.76	2.5	3.4	2.0
P 470	13.35	4.70	3.35	3.4	0.45	2.8	3.6	2.2

[a]See references 2 and 30.

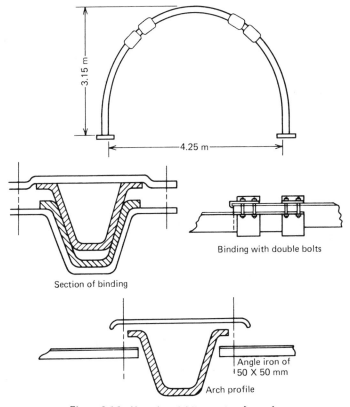

Binding with double bolts

Section of binding

Angle iron of
50 X 50 mm

Arch profile

Figure 2.16 Kunstler yielding arches [2, 26].

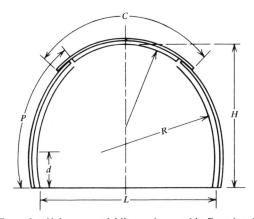

Figure 2.17 Toussaint–Heinzmann yielding arches used in French collieries [2, 30].

85

2.5.2   Estimation of Yielding Arches

Yielding arches, lowering 30–40 cm in height, cannot give a static model for calculations. The size estimation is done according to the convergence criteria of the roadway. The following formulas and tables make estimations of the conditions for yielding supports in German mines (31) [2, p. 195].

$$K = -78 + 0.666\,H + 4.3\,mK_t + 7.7\sqrt{10\,K_f} \qquad (2.28)$$

$$K' = -58 + 0.039\,H + 3.7\,mK_t + 6.6\sqrt{10\,K_f} \qquad (2.29)$$

$$Y = 3.5 + 0.23\,K$$

where K = final convergence, percent.
$\quad K'$ = heaving of the floor, percent
$\quad Y$ = closure of sides, percent
$\quad H$ = depth of the gateway, in meters
$\quad m$ = thickness of the seam, in meters
$\quad K_t$ = coefficient, according to the support of gateway ribs (Table 2.9).
$\quad K_f$ = coefficient according to the floor rock (Table 2.10).
$K'/K < 0.7$ yielding profiles 26–29 kg/m
$K'/K > 0.7$ yielding profiles 30–36 kg/m

As a numerical example, let us estimate the size of the Toussaint-Heinzmann arches to be used at a gateway at a depth of 1000 m driven in a seam of 2 m in thickness, made of sandstone floor, with

Table 2.9   Coefficient K_t, According to Support of Gateway Ribs[a]

Support of Gateway Ribs	K_t
Solidified materials like anhydrite or fluid concrete	1
Wooden chocks	2
Hand stowing	3

[a]See references 2 and 31.

Table 2.10 Coefficient K_f, According to
Roof Rock[a]

Roof Rock	K_f
Sandstone	1
Sandy shale	2
Shale	3
Very deformed rock	4
Coal	5
Coal + shale + deformed rock	6

[a]See references 2 and 31.

wooden chocks to support the ribs. In these conditions

$H = 1000$ m

$m = 2$ m

$K_t = 2$

$K_f = 1$

$$K = -78 + 0.066 \times 1000 + 4.3 \times 2 \times 2 + 7.7\sqrt{10 \times 1}$$
$$= -78 + 66 + 17.2 + 24.3 = 29.5\%$$
$$K' = -58 + 0.039 \times 1000 + 3.7 \times 2 \times 2 + 6.6\sqrt{10 \times 1}$$
$$= -58 + 39 + 14.8 + 20.9 = 16.7\%$$
$$\frac{K'}{K} = \frac{16.7}{29.5} = 0.56$$

To cover the convergence a Toussaint–Heinzmann profile of 21 kg/m should be used.

CHAPTER 3

Roof Bolts and Trusses

3.1 PRINCIPLE OF ROOF BOLTS

It is an established fact that there are tension zones in the roof, especially at the "entries" of the coal mines. The roofs in these entries act like beams supported on both sides with layers separated from each other. The designer of the roadway supports must take the weight of such separated beds (immediate roof) into consideration.

Let us consider two roof layers whose thicknesses are h_1 and h_2 and widths b (Fig. 3.1) [2, p. 482]. If the span of the opening is l and the uniform load is q, there will be a maximum bending stress in the middle (Fig. 3.1a) as follows:

$$\sigma = 0.75 \frac{ql^2}{bh_1^2 + bh_2^2} \qquad (3.1)$$

If these two layers are tied together by means of bolts (Fig. 3.1b), the bending σ' in the middle of the span would be

$$\sigma' = 0.75 \frac{ql^2}{b(h_1 + h_2)^2} \qquad (3.2)$$

It can be seen that the value of σ' is much less than σ. If $h_1 = h_2 = h_0$ the ratio of two cases is as follows:

$$\frac{\sigma}{\sigma'} = \frac{0.75(ql^2/2bh_0^2)}{0.75[ql^2/b(2h_0^2)^2]}$$

$$= 2$$

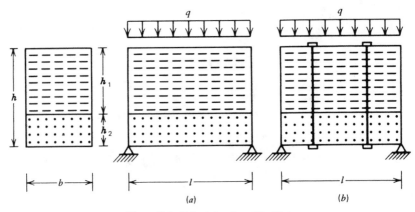

Figure 3.1 Principle of roof bolting.

Therefore, by binding the two layers, the bending stress can be reduced to half.

On the other hand, the tension stress met on the roof can be carried by "steel" rods quite resistant to tensile stresses. The binding of the layers can be effected as soon as the roadway is opened, without much bed separation. These obvious advantages have made roof bolting very popular in room-and-pillar workings in flat-bedded deposits. Investigations of the U.S. Bureau of Mines in this respect after World War II made a big improvement in roadway supports and made roof bolts very popular.

3.2 VARIETIES OF ROOF BOLTS

Among the varieties of roof bolts slot-and-wedge and expansion-shell bolts are anchored mechanically. In "grouted" bolts the setting medium is quick-setting cement. "Resin" bolting is the latest improvement, where varieties of quick setting resins are used quite efficiently in fixing the bolt in place.

3.2.1 Slot-and-Wedge Bolts

A typical slot-and-wedge bolt is shown in Fig. 3.2 [2, p. 485]. The bolt is made of malleable steel 22–30 mm in diameter and 0.5–2.5 m

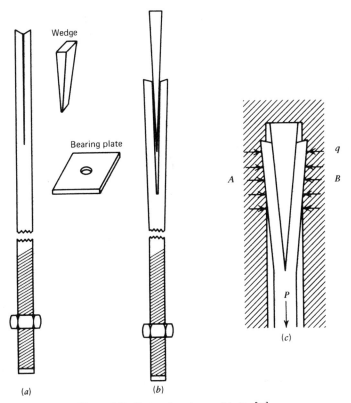

Figure 3.2 Slot-and-wedge roof bolts [2].

long. One end of the bolt is slotted (150 mm long, 2–3 mm wide), and a wedge of $\frac{1}{7}-\frac{1}{10}$ in conicity is placed and forced to enlarge the slot in place as shown in Fig. 3.2c.

The hole for the bolt is drilled according to the length of the bolt, usually 4 mm larger, and the bolt is forced to enlarge by a pneumatic impact hammer. The necessity for compressed air makes the usage quite inconvenient. After the bolt is fixed, the bearing plate is tightened by nuts, giving proper tension to the bolt. The anchorage force of the bolt is given by Cox [23] as follows:

$$P = F_t q \ (\sin \alpha + \mu \cos \alpha) \tag{3.3}$$

$$\mu = \kappa q \tag{3.4}$$

where P = anchorage force to keep the bolt in place, in kilograms

F_t = area of anchorage, in square centimeters

q = bearing capacity of roof rock, in kilograms per square centimeter

α = conical angle of the wedge

μ = coefficient of friction between roof rock and bolt steel

κ = coefficient, 0.0014

As a numerical example, let us use a rock with a bearing capacity of 200 kg/cm². A bolt with 2° of conicity and a 25-cm² area will have the following anchorage force:

$$\mu = 0.0014 \times 200 = 0.28$$

$$P = 25 \times 200 \ (\sin 2° + 0.28 \cos 2°)$$

$$= 25 \times 200(0.0349 + 0.2798) = 1573.5 \text{ kg}$$

3.2.2 Expansion-Shell Roof Bolts

A typical expansion-shell bolt is shown in Fig. 3.3. It consists of 17–22 mm steel rod that holds a conical piece N on the threaded end. There are four shells E around this central piece, moving horizontally as the N-piece moves downward with the help of a tightening wrench. The tightened part is seen in Fig. 3.3c where anchorage force is in equilibrium with the friction force S formed by qF_t forces:

$$P = n\mu q F_t \tag{3.5}$$

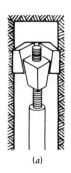
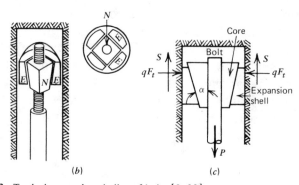

(a)   (b)   (c)

Figure 3.3 Typical expansion-shell roof bolts [2, 33].

where P = anchorage force, in kilograms

μ = coefficient of friction between rock and expansion shells

q = bearing capacity of roof rock, in kilograms per square centimeters

F_t = area of one expansion shell

n = number of shells

For a practical example, let us calculate the anchorage force of an expansion-shell bolt, 4 pieces, 5 cm² of friction surface of each, at a rock of 200 kg/cm² bearing capacity. The coefficient of friction is 0.28, as in the previous example.

$$P = 4 \times 0.28 \times 200 \text{ kg/cm}^2 \times 5 \text{ cm}^2$$

$$= 1120 \text{ kg}$$

The tightening action is schematically shown in Fig. 3.4 [2, p. 588], and the torque required is calculated by the help of the formulas [33, p. 399] in the following discussion.

As seen in Fig. 3.4, the nut S is turned by the help of wrench A, and the torque is read at G. By tightening, the expansion shells G_p

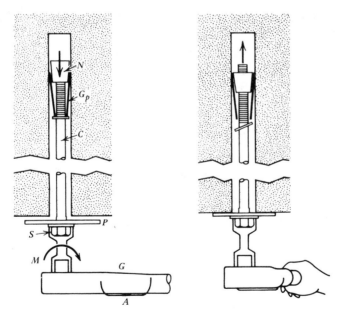

Figure 3.4 Tightening of expansion-shell roof bolts [2, 33].

moves horizontally, squeezing into the rock. The tightening continues with the bearing plate P until a pretension is given to the bolt. This should not exceed 60% of the flow stress of the steel used. The moments are as follows:

$$M = M_1 + M_2 = \frac{Rd}{2} \tan(i + \varphi_1) + \frac{R}{3} \frac{d_2^3 - d_1^3}{d_2^2 - d_1^2} \tan \varphi_2 \qquad (3.6)$$

where M = total turning moment, in kilogram-centimeters

M_1 = first moment put the shells into action, in kilogram-centimeters

M_2 = second moment to tighten bearing plate, in kilogram-centimeters

R = axial force applied to the bolt, in kilograms

d = diameter of the bolt, in centimeters

d_1 = diameter of the hole, in centimeters

d_2 = distance of expansion shell in the rock

i = inclination of bolt thread

φ_1 = angle of friction between nut and bolt

φ_2 = angle of friction between nut and bearing plate

As a numerical application, let us calculate the moment (torque) to get 10 t of force in a bolt 2.5 cm in diameter, put in a hole 3 cm in diameter. The distance of friction of shells is 4.5 cm; the angle of bolt thread is so small that it is negligible, and the angles of frictions are $\tan \varphi_1 = 0.2$ and $\tan \varphi_2 = 0.3$, between nut and bolt and nut and bearing plate respectively.

$$M = 10000 \frac{2.5}{2} \times 0.2 + \frac{10000}{3} \frac{(4.5)^3 - (3)^3}{(4.5)^2 - (3)^2} \times 0.3$$

$$= 8200 \text{ kg} \cdot \text{cm or } 82 \text{ kg} \cdot \text{m}$$

According to reference 2 the turning moment can be calculated in a simpler way as follows:

$$M = \frac{Rd}{2} (\tan i + 2 \tan \varphi) \qquad (3.7)$$

where M = torque, in kilogram-centimeters

R = axial force applied to the bolt, in kilograms

d = diameter of the bolt, in centimeters

i = inclination of bolt thread, usually taken as 2.5°
φ = angle of friction of nut on the bearing plate, usually taken as 16°

The foregoing calculation then becomes

$$M = \frac{10000 \times 2.5}{2} \ (\tan 2.5° + 2 \times \tan 16°)$$

$$= \frac{10000 \times 2.5}{2} \ (0.0437 + 2 \times 0.2867)$$

$$= 7715 \ \text{kg} \cdot \text{cm} = 77.15 \ \text{kg} \cdot \text{m}$$

which is in accordance with the result (82 kg · m) already calculated.

3.2.3   Grouted Roof Bolts

A schematic view of grouted bolts is given in Fig. 3.5, where grout (cement/fine sand/water) is put in to half of the length of the hole. A tap is used to stop the grout from running down. Fine plastic tubing is placed to drain the air while inserting the corrugated steel rod. After the cement sets, it has high adherence and keeps the bolt in place.

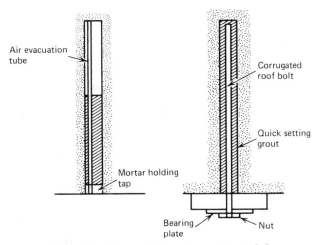

Figure 3.5   Schematic view of grouted bolts [2].

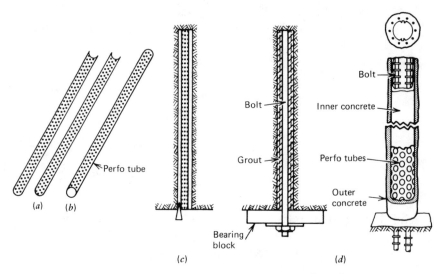

Figure 3.6 Grout sleeves used in fissured rocks [2, 34].

In fissured rock the water of the mortar is lost very easily. To eliminate this inconvenience, perforated sleeves are used as shown in Fig. 3.6 [34]. As the bolt is inserted, the mortar is forced through perforations of the sleeve to the hole.

The several ratios of cement to fine sand and water content (α = water/cement by weight) are seen in Fig. 3.7. It can be seen that strength increases with age, and half of the strength is reached in a week. Thicker grouts increase strength as well. The strength of grout is acutely dependent on the cement, sand, water ratios.

3.2.4   Resin Roof Bolts

Difficulties in fixing bolts have led to anchoring bolts along their full lengths. Grouted bolts were successful but needed long curing time and there was bed separation. It was also difficult to make proper mixture. So, some resins were developed that became hard and obtained mechanical properties in a few minutes.

Resin bolting, as it is generally called, is relatively new. The components of resin differ with different manufacturers. Different percentages of components will have different strengths, gel times, resistance to environment, and so on. Key components of a resin

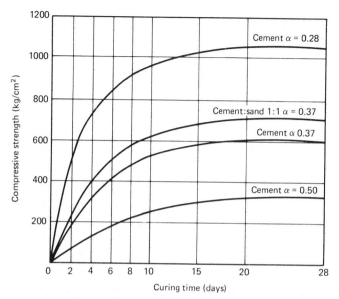

Figure 3.7 Curing time of mortar [2, 34].

bolt may typically be [29, p. 157] the following:

Polyster resin 28.5%
Filler (crushed limestone) 66% ⎬ + Catalyst
Accelerator 0.5%

Filler is any crushed rock. It is used to reduce shrinkage and also to reduce the amount of polyster resin to a minimum, because resin is much more expensive than rock. The accelerator assists the reaction between the catalyst and polyster resin so that the mixture cures faster. To avoid contact before use, resin, filler, and accelerator are packed together and separated from the catalyst. The common practice is to pack them in form of sausage cartridge, one packet inside the other, separated by a quality plastic wrap such as Mylar. Fig. 3.8

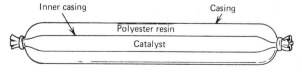

Figure 3.8 Resin cartridge manufactured by Du Pont [29, 35].

shows a typical arrangement of resin cartridges manufactured by
Du Pont [35]. The length of cartridge ranges from 30 to 120 cm,
diameter, 2.5 to 3.5 cm. Maximum anchoring capacity will be real-
ized in less than five minutes, and when completely cured the resin
has the following physical properties [36]:

Uniaxial compressive strength	1120 kg/cm^2
Tensile strength	630 kg/cm^2
Shear strength	525 kg/cm^2

When considering the load bearing capacity of a resin anchor, two
important points must be considered. First is rock strength (or better
referred to as rock type). It is widely appreciated that weaker rocks
require more resin to give anchoring characteristics comparable to
those achieved in stronger rocks. Second, bond length influences the
bolt's anchor strength. Fig. 3.9 gives the results of Franklin and
Woodfield [37], which show the interrelationship of these parameters. Their results generally indicate that the anchor strength is a
linear function to bond length (Fig. 3.9b). The rocks used to deter-
mine this relationship were granite, limestone, sandstone, coal, and
chalk and involved some 200 individual tests. The resin anchorage
strength proved to be 1.7–3 times greater than that of the mechanical type [36, p. 320].

The installation steps for resin bolting are summarized in Fig. 3.10
[39]: drill the hole, insert the cartridges, put the bolt through, turn
the bolt to get thorough mixing, and apply thrust with the proper
machine for 20–30 seconds [29, p. 159].

The bearing capacity of a resin bolt can be calculated as seen in
Fig. 3.11 and formulated as follows:

$$R_{max} = \sigma_a F = \tau U l \qquad (3.8)$$

$$F = \frac{\pi}{4} d^2, \qquad U = \pi d$$

$$\tau = 0.25 \frac{\sigma_a d}{l} \qquad (3.9)$$

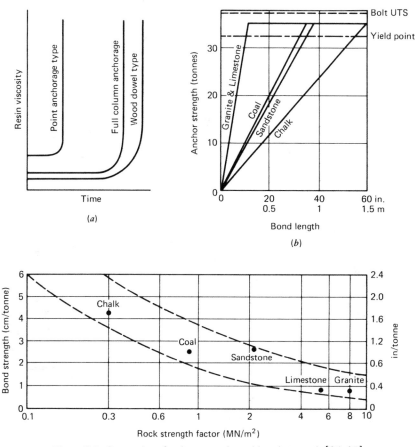

Figure 3.9 Properties of polyester resin and bond strength [36, 37].

where R_{max} = bearing capacity of bolt, in kilograms

σ_a = yield strength of bolt steel, in kilograms per square centimeter

F = area of the bolt, in square centimeters

d = diameter of the bolt, in centimeters

τ = adherence between resin and the bolt, in kilograms per square centimeter

U = circumference of the bolt in centimeters

l = length of the bolt, in centimeters

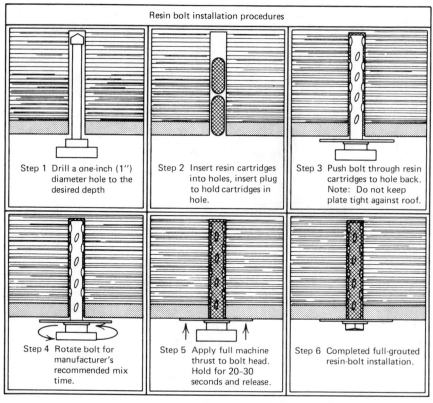

Figure 3.10 Resin bolt installation recommendations [29, 39].

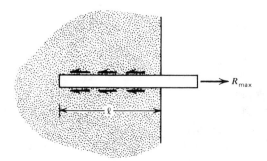

Figure 3.11 Bearing capacity of a resin bolt [2].

Let us take the yield strength of the bolt steel to be 2000 kg/cm^2, the diameter 2.5 cm, length 200 cm. Then the adherence between resin and steel and the bearing capacities of bolt is calculated as follows:

$$\tau = 0.25 \times \frac{2000 \times 2.5}{200} = 6.25 \text{ kg/cm}^2$$

$$R_{max} = \tau U l = 6.25 \times 2.5 \, \pi \times 200$$

$$= 9812.5 \text{ kg}$$

3.2.5 Wooden Roof Bolts

Wooden bolts, held in place by a resin column, are used as an anchorage system to strengthen cracked coal faces, coal ribs, and the like. A typical application is seen in Fig. 3.12 [44, 2, p. 504].

At the face, holes up to 16 m in length, are drilled, and wooden bolts, 36 mm in diameter, are placed. A 10-mm plastic tube is also placed to evacuate the air in the hole during the injection. The mouth of the hole is securely held by a plug, and the resin is injected by a pump at a pressure of 14–21 kg/cm^2, as seen in the figure.

Such a bolt increases the stability of the weak ground and increases the safety in mechanized faces. They do not cause any inconvenience, as the winning machines can easily cut them.

3.2.6 Testing of Roof Bolts

The bearing (anchoring) capacity is an important factor in bolt design. It is the load carried without any appreciable deformation. This capacity depends upon the roof conditions (strength of roof rock, fissures, etc.), atmospheric conditions (temperature, relative humidity), type of roof bolt (mechanical, grouted, resin, etc.), the method of anchorage of the bolt, and finally the strength of roof-bolt steel. This is obtained by in situ measurements in the mines.

A typical testing system is shown in Fig. 3.13 [38, 2, p. 512]. As seen in the figure the bolt (2) is pulled down by a hydraulic jacket (3), and the displacement is measured by an extensometer (4). The

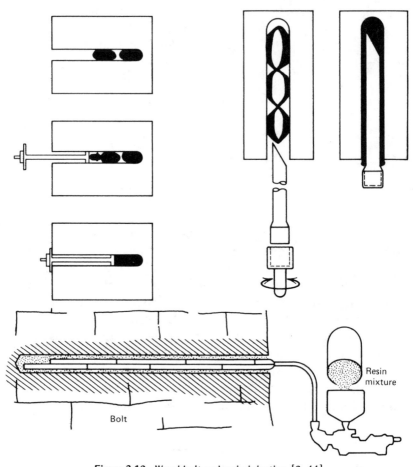

Figure 3.12 Wood bolt and resin injection [2, 44].

pressure exercised by a hand pump (6) is read at the manometer (5). The result of testing is seen in Fig. 3.14 (38).

This relation, as shown in Fig. 3.14, is linear, composed of two parts. There is no extension at the start even though there is a load applied (point 1). This is explained by the fact that the load applied has not yet reached the bolt. After the point 2, the linear extension increases (section 2–3), and the anchorage capacity is lost after point 3. The value at point 4 is the approximate working load of the bolt.

The anchorage capacities of various types of roof bolts are shown in Fig. 3.15 [39, 2, p. 514]. It can be seen that the resin bolt 25 mm

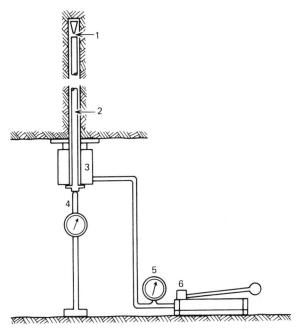

Figure 3.13 In situ testing of roof bolts [2, 38].

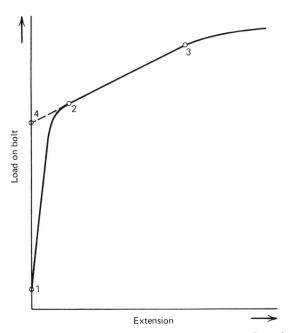

Figure 3.14 Characteristics of load-extension relations [2, 38].

103

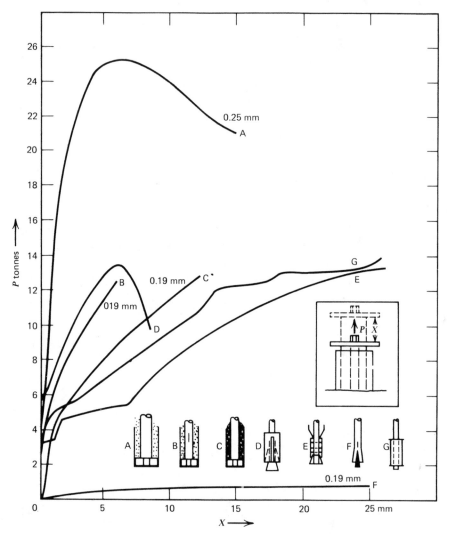

Figure 3.15 Anchorage capacity of various bolts [2, 39].

in diameter (A) shows the highest bearing capacity. Among the 19-mm bolts, resin ones show the steepest curves (B, C). Mechanical bolts show poor results (E, G), and the slot-wedge variety (F) is the poorest.

The factor of roof rock is clearly shown in Fig. 3.16 for expansion-shell bolts anchored in different rocks [17, p. 659]. Stronger rocks show higher anchorage capacity.

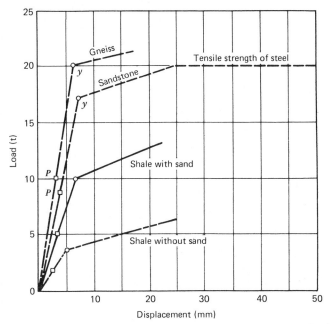

Figure 3.16 Admissible load on bolts as a function of rock quality [17].

3.3 DESIGN OF ROOF BOLTS

3.3.1 Stability of Bolted Blocks

Let us assume that a block is developed by two cracks at the side of a gallery at angle α to the horizontal (Fig. 3.17). The weight of such block is P. If the shear force along the crack surface exceeds the frictional force, the block moves, that is, there is caving of the block.

$$T_\alpha = P \sin \alpha \cdots N_\alpha = P \cos \alpha \qquad (3.10)$$

$$R_s = N_\alpha \tan \varphi \qquad (3.11)$$

$$\quad = P \cos \alpha \tan \varphi \qquad (3.12)$$

$$R_s \geqslant T_\alpha \qquad (3.13)$$

$$R_1 = P_c \cos \gamma = P_c \cos (\alpha + \beta) \qquad (3.14)$$

$$R_2 = P_c \sin \gamma \tan \varphi = P_c \sin (\alpha + \beta) \tan \varphi \qquad (3.15)$$

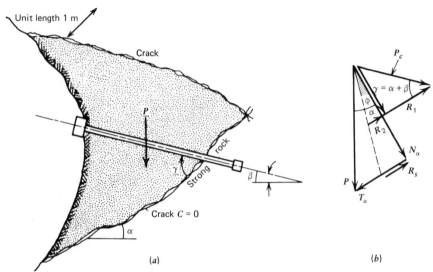

Figure 3.17 Carrying capacities of rock bolts [2].

$$n = \frac{\Sigma R}{T_\alpha} = \frac{R_s + R_1 + R_2}{T_\alpha} \tag{3.16}$$

$$= \frac{P \cos \alpha \tan \varphi + P_c [\cos (\alpha + \beta) + \sin (\alpha + \beta) \tan \varphi]}{P \sin \alpha} \tag{3.17}$$

$$P_c = \frac{(n \sin \alpha - \cos \alpha \tan \varphi) P}{\cos (\alpha + \beta) + \sin (\alpha + \beta) \tan \varphi} \tag{3.18}$$

where P = dead weight of the block separated by crack surfaces, in kilograms

α = angle of crack to the horizontal, in degrees

β = angle of the bolt to the horizontal, in degrees

φ = angle of friction at the crack surface, in degrees

R_s = friction force, in kilograms

P_c = axial force given to the bolt, in kilograms

n = factor of safety

ΣR = sum of forces against movement, in kilograms

N_α = normal force to the crack surface due to dead weight of the block

T_α = force making the movement, in kilograms

As the crack is supposed to be open, there would be no cohesion ($c = 0$) as seen in Eq. 3.11.

As a numerical example, let us find the tightening force with a safety factor $n = 2$ on a block separated by cracks making $\alpha = 60°$ angles and 1 m in length along a gallery height of $L = 1.5$ m. The angle of friction on crack surface is $\varphi = 25°$, and the inclination of the bolt to the horizontal is $\beta = 30°$. The density of rock is 2.5 t/m³. The dead weight of the block is as follows:

$$P \cong \tfrac{1}{2} L^2 \sin \alpha \cos \alpha \times 1 \times \gamma$$

$$\cong \tfrac{1}{2} (1.5 \text{ m})^2 \times 1 \text{ m} \times \sin 60 \cos 60 \times 2.5 \text{ t/m}^3$$

$$\cong 1.22 \text{ t}$$

$$P_c = \frac{(2 \sin 60° - \cos 60° \tan 25°)\, 1.22}{0 + 0.466}$$

$$= 3.92 \text{ t}$$

A bolt with larger axial force can easily hold such a block in place. The block will move if a bolt of a lower force is used.

3.3.2 Length of Bolts

According to investigators [40, 41], the length of the bolts l should be greater than the dome height separated from the main roof. If the gallery width is L, these lengths are as follows:

$$\text{Strong roofs} \qquad l = \tfrac{1}{3} L \qquad\qquad (3.19)$$

$$\text{Weak roofs} \qquad l = \tfrac{1}{2} L \qquad\qquad (3.20)$$

For very strong roofs where the bolting is done to stop spalling, the length is $l = 1$ m, as a minimum.

3.3.3 Spacing of Bolts

The spacing of bolts is closely related to the length of the bolt. According to the photoelastic investigations of Coates and Cochrane [42] the spacing should be as follows:

$$b = \tfrac{2}{3} l = \tfrac{2}{9} L \qquad\qquad (3.21)$$

$$l_{max} = \frac{R_{max}}{b^2 \gamma} \tag{3.22}$$

where b = spacing of bolts, in meters

L = width of gallery, in meters

l = length of bolt, in meters

R_{max} = maximum carrying capacity of bolt; the force resulting in yield in the steel, in tonnes

γ = density of rock, in tonnes per cubic meter

If the tightening of the bolt is less than 0.5 σ_{max} of the steel strength, the spacing should be taken as half of this value [43].

3.3.4   Diameter of Bolts

The diameter of bolts is calculated according to the yield strength of the steel.

$$R_{max} = \sigma_a F \tag{3.23}$$

$$R = \frac{R_{max}}{n} = \frac{0.785 \, d^2 \sigma_a}{n} \tag{3.24}$$

where R_{max} = maximum bearing capacity of bolt (in tension), in kilograms

R = allowable axial force in bolt, in kilograms

n = safety factor, 2–4

σ_a = yield strength of steel, in kilograms per square centimeter

F = area of the bolt, in square centimeters

d = diameter of the bolt, in centimeters

Figure 3.18 shows carrying capacity of roof bolts with a safety factor of 2 for steels St. 37 and St. 52 [2, p. 526]. The practical diameters for shale, limestone, and sandstone are 30, 35, and 40 mm, respectively.

3.3.5   Density of Bolts

The number of bolts per square meter is called "density." It is a usual practical to have this number be 1. In poor, fractured roofs the density is increased.

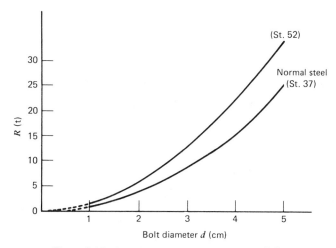

Figure 3.18 Carrying capacities of roof bolts [2].

3.3.6 Numerical Example

The foregoing design criteria of bolts can be applied to a numerical example. The data available are as following and are shown in Fig. 3.19 [2, p. 528]:

Width of gallery	$L = 3$ m
Roof conditions	fractured
Roof rock	coal
Immediate roof thickness	$h = 1.75$ m
Immediate roof density	$\gamma = 2.5$ t/m^3
Distance between rows of bolts	$c = 1$ m

The length, according to Eq. (3.21) is

$$l = \frac{L}{2} = \frac{3}{2} = 1.5 \text{ m}$$

As there is an immediate roof to be separated from the main roof, the length should exceed this thickness by at least 0.5 m. So the length is

$$l = h + 0.5 = 1.75 + 0.5 = 2.25 \text{ m} \tag{3.25}$$

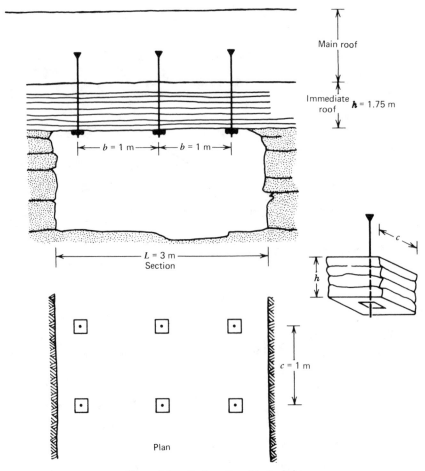

Figure 3.19   Design of roof bolts [2].

The bolts should carry the static weight of the roof shown in Fig. 3.19. The number m of the bolts is calculated as follows:

$$mR \geqslant Lhc\gamma \tag{3.26}$$

$$m \geqslant \frac{Lhc\gamma}{R} \geqslant \frac{Lhc\gamma n}{0.785\, \sigma_a d^2} \tag{3.27}$$

If we take $d = 2.5$ cm bolts of St. 37 ($\sigma_a = 2400$ kg/cm²) with a safety factor of $n = 2$,

$$m = \frac{3 \text{ m} \times 1.75 \text{ m} \times 1.0 \text{ m} \times 2.5 \text{ t/m}^3 \times 2}{0.785 \times 24000 \text{ t/m}^2 \times (0.025)^2 \text{ m}^2}$$

$$= 2.2 \sim 3$$

Bolt density m_0 is calculated as follows:

$$m_0 = \frac{m}{Lc} = \frac{3}{3 \times 1} = 1 \text{ piece per square meter}$$

Spacing of the bolts is determined

$$b = \frac{m_0}{c} = \frac{1}{1} = 1 \text{ m}$$

We should verify the length of the bolt with respect to maximum load as follows:

$$R_{max} = 0.785 \, \sigma_a \, d^2 = 0.785 \times 24000 \times (0.025)^2$$

$$= 11.775 \text{ t}$$

$$l_{max} \leqslant \frac{R_{max}}{b^2 \gamma} = \frac{11.775 \text{ t}}{(1.0)^2 \text{ m}^2 \times 2.5 \text{ t/m}^3} = 4.71 \text{ m}$$

Since the actual length $l = 2.25$ is smaller than 4.71 m, the length is quite safe.

3.4   APPLICATION OF ROOF BOLTS

Roof bolts are extensively used to support entries in room-and-pillar workings and gateways of longwalls, occasionally used in longwalls and in tunneling, and to some extent in the stoping of metal mines.

3.4.1   Room Entries

Roof bolts are most advantageously used in the entries of room-and-pillar workings. In the United States the bolts are put at 1.2-m intervals. Two examples are shown in Fig. 3.20, one holding together the shaly roof layers and the other hanging the coal to the shaly roof.

At the junctions, where the span is greater, extra bolts are put at greater depth, as shown in Fig. 3.21 [44, 2, p. 532].

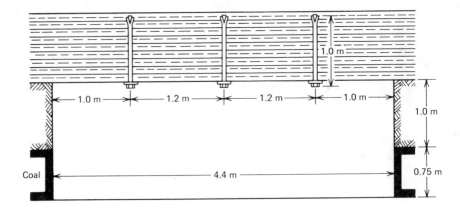

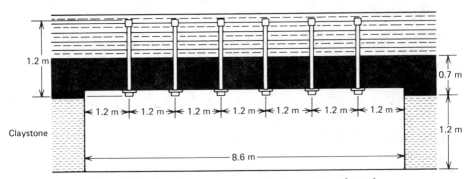

Figure 3.20 Roof bolting in room-and-pillar workings [2, 44].

3.4.2 Gateways of Longwalls

The uses of resin bolts in a gateway in the British coalfields are shown in Fig. 3.22 [39, 2]. As shown in the figure, the roof of the gateway is held in place by four resin bolts held together by U-bar Fig. 3.22a. The bolts are 1.63 m long. A wood bolt is placed in the coal on the rib side of the gateway as well. The roadway convergence is shown in Fig. 3.22c where 1.5-m convergence decreased to 0.25 m with roof bolting. Roof control is increased by the use of roof bolting [39].

An interesting application is the use of wooden roof bolts to counteract heaving of the floor, as illustrated in Fig. 3.23 [45, 2, p. 535]. Resin bolts reduced heaving tremendously when placed both

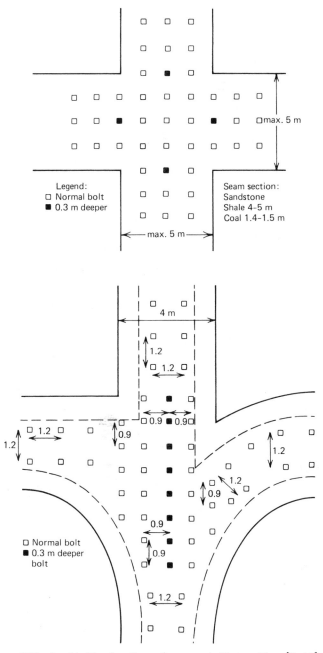

Figure 3.21 Roof bolting junctions of room-and-pillar workings [2, 44].

113

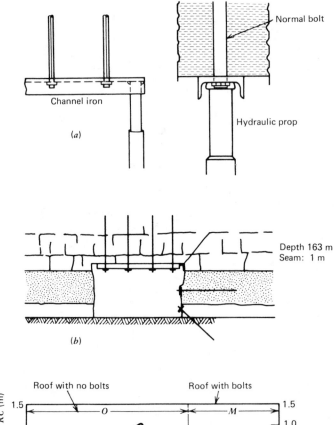

Figure 3.22   Roof bolting in gateways of longwall workings [2, 39].

vertically and at a 45° angle to strengthen the floor rock. This system eliminated the previously used alteration of arches.

3.4.3   Longwall Faces

Figure 3.24 shows many applications of wooden bolts to longwall faces [39, 2, p. 538]. Figure 3.24*a* shows 4–5-m wooden bolts to

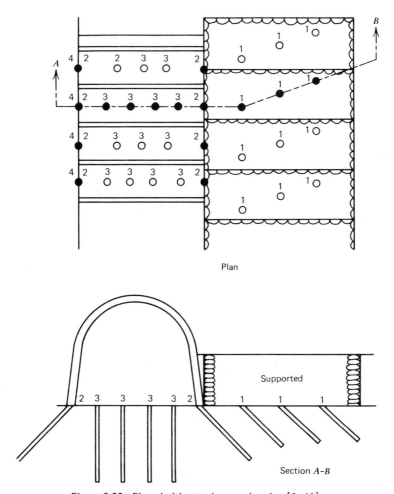

Plan

Section A–B

Figure 3.23 Floor bolting to decrease heaving [2, 45].

strengthen a weak but thick seam. Figure 3.24*b* illustrates the strengthening of a fault zone in front of the face by 15-m wooden bolts with resin cementing. Figure 3.24*c* shows wooden bolts driven into the roof to tie fractures produced under a pillar left in the upper seam. Such small pillars, left, are under high abutment pressures resulting in fractures of the roof and floor rocks. These wooden bolts do not cause any trouble during mechanical winning of the coal.

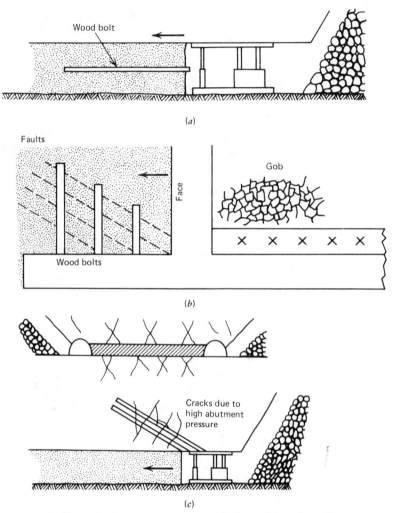

Figure 3.24 Wooden bolting used at longwall faces [2, 39].

3.4.4   Metal Ore Mines

Arch roadways, driven unsupported into strong rock, may spall, and dangerous roof falls may occur. To eliminate this spalling of gallery roof and sides, roof bolting may be quite satisfactory (Fig. 3.25).

In stoping, especially hydraulic cut-and-fill stoping, the hanging

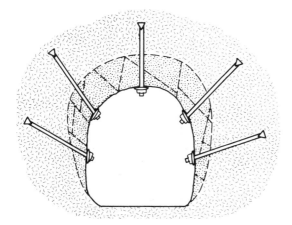

Figure 3.25 Bolting against spalling [2].

wall can be supported by roof bolts, which eliminates any cavings and dilutions of the ore Fig. 3.26 [46, 2, p. 540].

The bolting of a fault zone is illustrated in Fig. 3.27, where the deformation of the side piece of the arch was eliminated by such bolting [46, 2, p. 541].

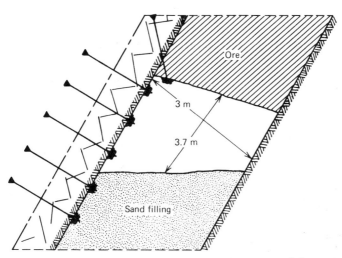

Figure 3.26 Bolting in hydraulic cut-and-fill stopping [2].

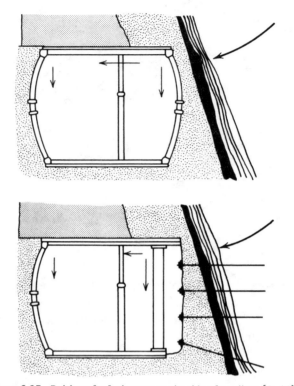

Figure 3.27   Bolting of a fault zone on the side of a gallery [2, 46].

3.5   ADVANTAGES OF ROOF BOLTING

The advantages of roof bolting over other supporting systems can be summarized as follows:

1.   Bolts can be put in as soon as the excavation is made, before any appreciable deformations. This is the most important factor in roof bolting, helping the roof control, increasing the safety.

2.   Bolts, especially resin bolts, are not influenced by the shock waves of explosives.

3.   There are no posts, girders, and the like to obstruct the galleries. The hauling equipment can easily pass. The cross section is kept open.

4. The resistance to the air flow is low; ventilation is improved.
5. Bolts are natural supports to hang pipes, tubes, and the like and thus clear the floor for free passage.
6. Spalling is reduced to quite an extent, reducing dilution of coal from the falling roof rock.
7. Roof bolting is more economical then other support systems. In many mines, where timber is not obtained freely and cheaply, roof bolting is much less costly, and there is no capital expenditure for gallery steel arches.

3.6   ROOF TRUSSES

3.6.1   Principle and History of Roof Trusses

Roof trusses were designed according to the patent of White [47] and his improvements [48, 49]. Full-column resin bolts had replaced all other kinds of bolting because of lower cost. However, where the ground is heavy roof trusses are the remedy to hold places otherwise only held by timbering [49].

It is a well-known fact that flat-roofed openings develop tension zones at the roof. As the trusses put stress to the roof, these tension zones are eliminated.

The general view of a roof truss is shown in Fig. 3.28 [49]. It consists of two-point anchoring system (preferably resin), a connection bar, a turnbuckle to give the proper tension to the bar, bearing blocks, and adjusting wedge box.

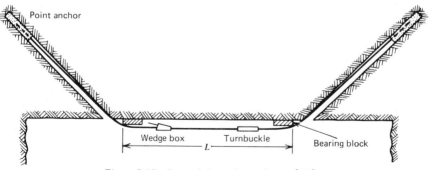

Figure 3.28   General view of a roof truss [49].

3.6.2   Design of Roof Trusses

A roof truss is detailed in Fig. 3.29 [50]. The tension P on the bar is given by the turnbuckle T. Through the bearing block $C\text{-}D$ ($2a \times b$) reactions R_2 are formed, and through the touching of the hole mouth reactions R_1 are formed.

Resolving the forces along and perpendicular to the direction of T, and taking moments around point B, we find:

$$T - \mu R_2 - R_1 \sin \alpha - P \cos \alpha = 0 \tag{3.28}$$

$$R_2 + R_1 \cos \alpha - P \sin \alpha = 0 \tag{3.29}$$

$$R_2(a + l) + \mu R_2 b - Tb = 0 \tag{3.30}$$

If we solve these equations simultaneously in respect to T,

$$P = \frac{T}{\mu b + a + l} [(a + l) \cos \alpha + b \sin \alpha] \tag{3.30}$$

$$R_1 = \frac{T}{\mu b + a + l} [(a + l) \sin \alpha - b \cos \alpha] \tag{3.31}$$

$$R_2 = \frac{Tb}{\mu b + a + l} \tag{3.32}$$

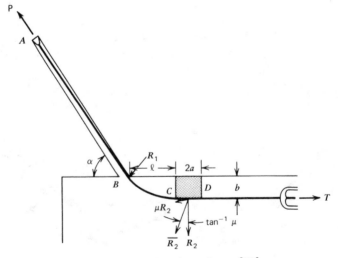

Figure 3.29   Statics of a roof truss [50].

In cases where block C-D is close enough to the hole that the rod does not touch the hole, $R_1 = 0$, then

$$T - \mu R_2 = P \cos \alpha$$

$$R_2 = P \sin \alpha$$

$$P = \frac{T}{\cos \alpha + \mu \sin \alpha} \tag{3.33}$$

$$R_2 = \frac{T}{\mu + \cos \alpha} \tag{3.34}$$

$$\alpha = \tan^{-1} \frac{b}{a+l} \tag{3.35}$$

where P = load on the anchorage of the bolt
T = tensioning load on the truss
R_1 = reaction at the mouth of the hole
R_2 = reaction at the block
l = distance of the block to the hole
$2a$ = width of the block
b = thickness of the block
μ = coefficient of friction between the block and the roof rock
α = angle of inclination of the hole

As a numerical example let us calculate the fixing force of a rod in the hole of a roof truss tightened at a load of $T = 10$ t. The holes are driven at an angle of $\alpha = 60°$, the block $b = 8$ cm thick and $2a = 20$ cm long is placed at a distance of $l = 22$ cm from the hole. The coefficient of friction between block and roof rock is $\mu = 0.4$. Find the angle and anchorage force if the same block is placed at 5 cm from the hole.

$$P = \frac{10,000}{0.4 \times 8 + 10 + 22} \, [(10 + 22) \cos 60 + 8 \sin 60)]$$

$$= \frac{10,000}{35.2} \times 22.93$$

$$= 6514 \text{ kg}$$

$$\alpha = \tan^{-1} \frac{8}{10 + 5} = 28°$$

$$P = \frac{10,000}{\cos 28° + 0.4 \sin 28°} = \frac{10,000}{0.8829 + 0.4 \times 0.4695}$$

$$= \frac{10,000}{1.0707} = 9340 \text{ kg}$$

It can be seen that a higher degree of anchorage is needed for this special case.

The reactions and anchorage force are plotted in multiples of T tension against the inclination of the hole as shown in Fig. 3.30 [50].

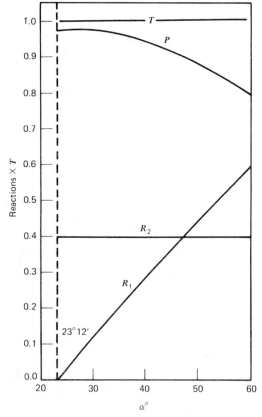

Figure 3.30 Anchorage reactions of a roof truss in terms of tension and angle of inclination [50].

Table 3.1 Dimensions for Good Roof Trusses[a]

Span Hole-to-Hole (m)	Block Thickness b (cm)	Block Distance to Hole l (cm)
2.6	8	20–22
3.0	8	20–22
3.6	10	25–30

[a] See reference 50.

To get good results in roof trusses, the angle of inclination of holes should be near 60°, the width of block $2a$ should be 20 cm, and the thickness and distance to the hole should be as listed in Table 3.1, according to the span (distance from hole to hole) of the gallery.

CHAPTER 4

Steel Longwall Supports

4.1 EVOLUTION OF STEEL LONGWALL SUPPORTS

The low strength and high cost of timber and the mechanization of longwalls led to the use of steel for support systems just before World War II, and this advancement led to the total mechanization and automation in longwalls, with high production and concentration of working places in the mines.

The early steel supports with friction props and articulated caps are shown in Fig. 4.1 [2, p. 544]. The plan view of an advancing longwall is shown in Fig. 4.1a, the positions of the steel supports in plan view at Fig. 4.1b, and the section view at Fig. 4.1c. A steel set is composed of a "prop" (1) and a "cap" (2) arranged in "T-form." Wooden wedges (3) may be placed according to the roof conditions. The set can be easily placed by setting a lock system (7) and can also be easily released by the same lock system (6) having the "back row" set transferred to the "face row," as shown by dotted lines in the figure. Thus the sets change place from back to front, not requiring any new support. The back of the face is "caved" in this advancement.

The articulation of caps permits the placement of the prop toward the end of the working shift thus creating an area of "prop-free face" for the chain conveyor (4) to move freely and for the coal winning machine (5) to cut and transfer the coal to the conveyor.

The props are telescopic, made of two pieces gliding one inside the

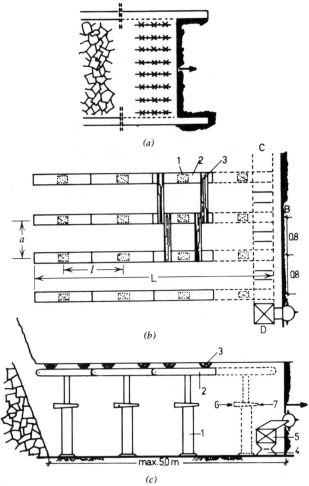

Figure 4.1 Steel longwall supports with friction props and articulated caps [2].

other, and set according to the thickness of the seam. The setting of the props is accomplished either by lock mechanism in friction props or by hydraulic mechanism or pressured fluid obtained from hoses at the face in hydraulic props. The caps are simple I-beams having an articulation and a setting mechanism for the articulation to support the roof for a short time.

Hydraulic props were further improved by making the hydraulic prop, cap, and the chain conveyor movement in a single unit called "walking support" or "powered chocks," as shown in Fig. 4.2 [51, 2, p. 544]. Schematically each unit of powered support consists of four to six legs (1) supported by large cap (canopy) (3). The winning machine (2) is placed on the chain conveyor and is always pushed to the face by the powered chock making "snaking," positions, as shown by the dotted line in the plan view. The supports self-advance by pulling action toward the chain conveyor, thus "walking" and making the roof "cave" as the advance is accomplished. There are several advances (three to six times) during the shift, according to the cuts performed, making an advance of 1.5–5 m by the winning machine and producing large tonnages with few workers. The recent improvements in powered supports, such as the "shield support," have made the back of the face much safer.

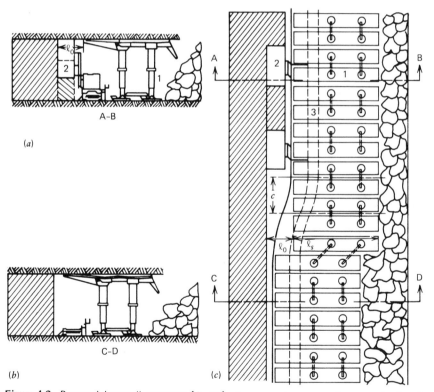

Figure 4.2 Powered longwall supports [2, 51]: 1, powered support; 2, cutter loader; 3, chain conveyor.

4.2   STEEL PROPS AND CAPS

4.2.1   Friction Props

The construction and working principle of friction props are shown in Fig. 4.3 [2, p. 545]. The prop is made of a cylindrical outside piece F and an inner piece P, connected by "wearing" plates a, and held by horizontal force H. This force is calculated as follows:

$$R = nH \tan \varphi \qquad \text{for cylindrical props} \qquad (4.1)$$

$$R = nH \tan (\varphi + i) \qquad \text{for conical props} \qquad (4.2)$$

where H = horizontal lock force, in tonnes
    R = yielding load, in tonnes
    φ = angle of friction between inner piece and bearing plates, $\varphi = \tan^{-1} 0.3\text{--}0.5$

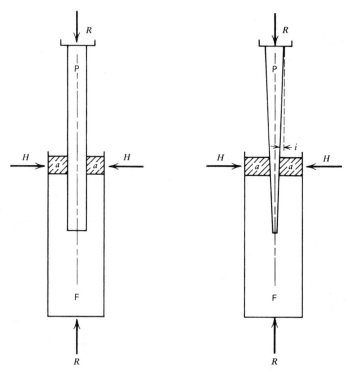

Figure 4.3 Principle of friction props [2].

i = angle of conicity of the inner piece, $i = \tan^{-1} 0.01$–0.005
n = number of friction surfaces, $n = 2$ in most of props

There are usually two friction surfaces, but there are props with more surfaces (slit props $n = 4$) and a variety of ways to increase the friction surfaces [7].

The working conditions of conical props and the characteristic curve (load versus sinking) are shown in Fig. 4.4a. These are called "slowly" loaded props, as the load carried is proportional to the sinking. As the inner piece sinks, the conicity of the inner piece forces the locking system to enlarge and exercise higher horizontal force on the locking system.

In cylindrical props where $i = 0$, the horizontal force is exercised

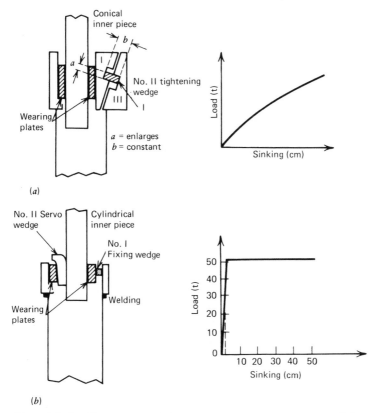

Figure 4.4 Locking systems and characteristic curves of friction props [2].

by an extra wedge called a "servo," as shown by II in Fig. 4.4b. Owing to high conicity ($\frac{1}{10}$), the servo exercises large locking force by sinking slightly (20 mm). For this reason these are called "instantaneous" loading props, as seen in the characteristical curve.

The profiles of friction props are usually rectangular, made of two corners or two channels welded together, as shown in Fig. 4.5. The weights and section moduli are also incorporated. It can be seen that owing to rectangular shapes, W_x and W_y are about equal, meaning that they have about the same strength for deformations in both directions [30, p. 402].

4.2.2 Hydraulic Props

The effects on the use of friction props caused by human errors in giving friction force, aging of the friction surfaces, and other difficulties made it necessary to design a better prop working on hydraulic principles, as shown in Fig. 4.6 [30, p. 426].

In Fig. 4.6a, opening and closing of valve 1 causes fluid to change place between the inner and outer pieces, lowering and stopping of props according to the load coming from roofs. It is so arranged that the prop stays at a load of 20 t (± 0.5 t), as shown in the characteristic curve of Fig. 4.6c.

The prop is raised and tightened by the help of a hand pump built into the prop, as schematically shown in Fig. 4.6b. By turning the handle along arc h, the piston moves up, opening valve 2, and lets some fluid pass from the inner piece to the outer piece, thus raising the prop. This can be done by adding outside pressurized fluid as shown in Fig. 4.6d. This eliminates the hand pump and makes the prop lighter, but it necessitates pressurized fluid hoses at the face. Finally, the prop can be lowered easily by opening valve v by pulling a chord at F. This lets the fluid flow from outer piece to the inner piece, lowering the prop. Hydraulic props work much better than friction props with easy setting and lowering, and they keep loads at the desired level, which causes less convergence.

4.2.3 Articulated Caps

The roof is held by articulated caps placed over friction or hydraulic props. Caps are iron beams, 1.0–1.25 m long, having an articulation

		Inner			Outer		
Type		Weight (kg)	W_x (cm^3)	W_y (cm^3)	Weight (kg)	W_x (cm^3)	W_y (cm^3)
1. Schwarz universal	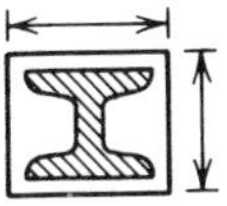	21	51	26	22	82	77
2. Schwarz radiac		13			16		
3. Semi	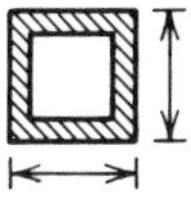	26	68	62	21	81	75
4. Becorit heavy model	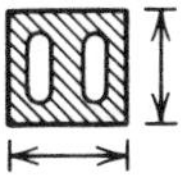	22	51	40	22	82	76
5. Gerlach duplex light		12	20	15	8	42	39
6. Gerlach duplex middle		16	34	22	10	60	53
7. Gerlach duplex heavy		20	51	34	13	84	86
8. Gerlach duplex extra heavy		25	70	46	16	84	86

Figure 4.5 Profiles and section moduli of friction props (30).

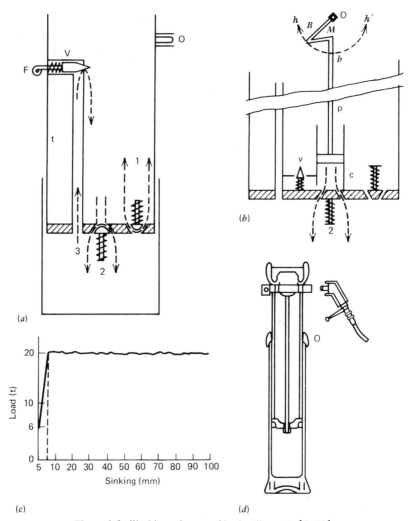

Figure 4.6 Working schemes of hydraulic props [2, 30].

where they are hooked together and stayed without a support for a short time. Such an articulation is secured by several wedged connections. A schematic view is shown in Fig. 4.7a [30, p. 443]. The articulation provides a "prop-free" distance of 2 m where the conveyor, winning machines, and pushers of the conveyors can operate freely. Such an articulation should hold blocks formed by 45° fractures which usually weigh 1–1.5 t as shown in Fig. 4.7b. The profiles of

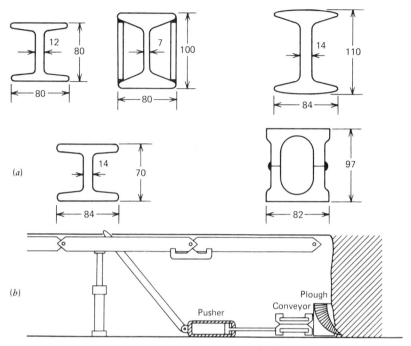

Figure 4.7 Articulated caps [30].

Figure 4.8 Typical prop-free front with articulated caps [30].

the caps are H-beams fortified by adding side pieces, turning them into square profiles. Channel irons are also welded together in square profiles, providing almost equal moduli of W_x and W_y.

The view of a longwall face with hydraulic props, articulated caps, and prop-free front is shown in Fig. 4.8 [30].

4.3   DESIGN OF PROPS AND CAPS

The design of props and caps includes provision for the density of props (number of props per square meter face area), the size of cap profile, and intrusion of the floor rock.

4.3.1   Prop Density Calculations

To calculate the requirements for prop density, the stresses are evaluated by the different formulas mentioned in Section 1.3.3 on wood supports. The dimensions are shown in Fig. 4.1b.

$$\sigma_t La = \frac{P_n kN}{n} \tag{4.3}$$

$$D = \frac{N}{La} \tag{4.4}$$

where σ_t = roof pressure evaluated, in tonnes per square meter
L = width of the face, distance held by supports, in meters
a = distance between rows of supports, in meters
P_n = nominal load of a prop, in tonnes
k = efficiency factor of props (Table 4.1)

Table 4.1.   Prop Efficiency Factor[a]

Type of Prop	Efficiency Factor k
40 t Friction	0.45
40 t Hydraulic	0.82
30 t Hydraulic	0.89
20 t Hydraulic	0.92

[a] See reference 2.

N = number of props per row

n = safety factor, usually 2

D = prop density, pieces per square meter

Let us calculate the distance between rows a and the prop density at a face of 2-m coal thickness supported by four friction props in a row of 40 t nominal capacity with 1.25-m articulated caps. The angle of internal friction of roof is $\varphi = 40°$ and the density of the roof rock is $\gamma = 2.5$ t/m³.

The roof pressure at the face, according to the Terzaghi formula, Eqs. (1.34), (1, 35), is

$$\sigma_t = \frac{B\gamma}{\tan \varphi}$$

$$B = B_1 + m \tan \left(45° - \frac{\varphi}{2}\right)$$

$$B_1 = \frac{L}{2} = \frac{5.0}{2} = 2.5 \text{ m} \quad \text{(in our case)}$$

where $m = 2$ m   (seam thickness)

φ = angle of internal friction, 40°

$$B = 2.5 + 2 \tan \left(45° - \frac{40°}{2}\right)$$

$$= 2.5 + 2 \tan 25°$$

$$= 3.43 \text{ m}$$

$$\sigma_t = \frac{2.5 \text{ t/m}^3 \times 3.43 \text{ m}}{\tan 40°}$$

$$= 10.22 \text{ t/m}^2$$

$$10.22 \text{ t/m}^2 \times 5 \text{ m} \times a = \frac{40 \text{ t} \times 0.45 \times 4}{2}$$

$$a = \frac{40 \times 0.45 \times 4}{2 \times 10.22 \times 5}$$

$$= 0.70 \text{ m}$$

$$D = \frac{4}{5 \times 0.70} = 1.14 \text{ per square meter}$$

4.3.2   Intrusion of Props

The floor rock of the seam should be able to stand a load without intrusion. Intrusion causes a large convergence and the trouble of moving the prop from the back of the face to the front. If the area of the prop is F and the safe bearing strength of rock is σ_{sf}, the stress developed:

$$\sigma = \frac{P_n k}{F} \leqslant \sigma_{sf} \tag{4.5}$$

If the safe bearing strength of floor rock is 40 kg/cm^2, and the dimension of the outer piece of the prop is 20 × 20 cm, the friction prop of 40 t nominal capacity will have the following bearing stress:

$$\sigma = \frac{40000 \times 0.45}{20 \times 20}$$

$$= 45 \text{ kg/cm}^2$$

which is more than the bearing strength of the floor rock. In such case larger size of prop is used, or the floor is dug further to get stronger rock, or larger additions are used as shown in Fig. 4.9.

4.3.3   Size of Caps

The caps tied together may be taken as a continuous beam supported by props, and the bending formula of Fig. 1.36 can be used.

$$\sigma = \frac{M_{\max}}{W} \leqslant \sigma_{sf} \tag{4.6}$$

For this example, where the face is supported by four caps and posts in a row, the maximum bending stress is

$$\sigma = \frac{0.1\sigma_t \cdot a \cdot (l)^2}{W} \leqslant 1400 \text{ kg/cm}^2$$

where σ_t = roof pressure = 10.22 t/m^2 = 1.022 kg/cm^2,
  a = distance between rows 70 cm,
  l = the span (distance between posts) 125 cm,

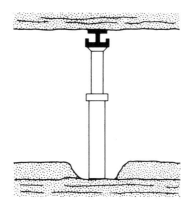

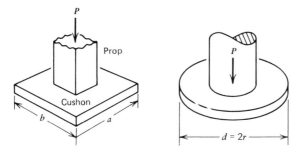

Figure 4.9 Precautions against floor intrusion [2].

If we take an I-beam GI-90 with $W_x = 62.5$ cm^3 (Table 2.3),

$$\sigma = \frac{0.1 \times 1.022 \text{ kg/cm}^2 \times 70 \text{ cm} \times (125 \text{ cm})^2}{62.5 \text{ cm}^3}$$

$$= 1788.5 \text{ kg/cm}^2$$

which is greater than the 1400 kg/cm^2 allowable stress in steel. Taking the larger size, that is, GI-100, $W_x = 80.7$ cm^3, we have the following:

$$\sigma = \frac{0.1 \times 1.022 \times 70 \times (125)^2}{80.7} = 1385 \text{ kg/cm}^2$$

which is safe. Any profile with larger W_x may be used for caps.

4.4   POWERED SUPPORTS

4.4.1   Development of Powered Supports

Powered supports have come after a long development of steel supports in longwall faces. Until World War II, friction props and bars were in use. Hydraulic props were developed in an effort to overcome defects from the aging of the friction surfaces, and human errors in preloading of the props. Convergence was decreased by the hydraulic working of the prop, but intrusions to the floor and intermittent changes from back to front of the face did not keep up with the pace of mechanical winning of the coal. Machines were developed swift enough to cut three to four times per shift, and the support changes could not keep up with this fast advance. A new system was developed, hydraulic in design, with props and caps incorporated into one unit and connected to the armoured conveyors to advance regularly with the cutting at the face line. Such support systems were named "walking chocks," as they advance (walk) by pulling themselves toward the conveyor. This system has been further improved in different designs, that make the back of the face safer with "shield" supports. Thus the hand output per manshift has been raised from 1.5 to 5 t, the face output per manshift (OMS), from 3.5 to 8 t. In England usage of the powered supports in longwalls has increased from nothing to almost 90% at present. Similar advancement can be seen in Germany, France, Poland, the USSR, and other European countries. High output per manshift, 100% recovery of coal, and limitations in pillared workings have led to their use in the United States. In 1976 about 4.6% of the production came from powered longwalls [29].

The development of powered supports is summarized in Table 4.2. The conditions met are shown by + and those not met by -. It can be seen that in powered supports all the conditions at the face are met.

4.4.2   Types of Powered Supports

Powered supports have been improved considerably since they were first made. Today there are chock, frame, shield, and chock-shield types of powered supports designed for various conditions. Only a general view can be given.

Table 4.2. Conditions to be Met at Longwall Faces[a]

Conditions	Friction Props	Hydraulic Props	Rigid Chocks	Powered Supports
Capability to meet pressure	−	+	+	+
Intrusions of the floor rock	−	−	+	+
Preloading, human factors	−	+	−	+
Keep up with coal cutting	−	−	−	+

[a] See reference 2.

The earliest powered support, manufactured by the Gullick Company of England, is a hydraulic chock as shown in Fig. 4.10a [2, p. 599]. The earliest model is composed of a block (chock) of one horizontal and four vertical pistons Fig. 4.10a. The pistons are connected to the caps of channel iron section, which are strengthened by iron sheets. The vertical pistons support the roof, and the horizontal piston pushes the conveyor. A typical chock support with six legs is shown in Fig. 4.10b [52, 29, p. 234], with different elements numbered and explained. The back of the chock is protected from caving blocks (8) and the rigid canopy (cap) has elongations to cover the roof after the cutting machine passed.

The frame support, first developed by the Dowty Company of England, is shown in Fig. 4.11a [2, p. 602]. The system is composed of two different supports of three- and two-leg units. The two-leg unit is connected to the conveyor and advances with the cutting machines to cover the freshly opened area of the root (Fig. 4.11b). The three-leg pieces are used to support the back of the face; after the cutting machine has passed, they advance and align again with the two-leg units. The modern support, made of two 2-leg units, is shown in Fig. 4.11c with the elements numbered. The canopy is articulated and has prolongations to cover the face area right after the cut.

Most recently, to keep up with easily caving faces, shield supports have been developed. These consist of an inclined plate whose lower end is hinged to a horizontal base plate that sits on the floor, while the upper end is hinged to a horizontal roof canopy in contact with the roof. The "caliper" shield has a single joint between the base and the gob shield such that when the hydraulic cylinders extend, the tip of the roof canopy moves away from the face line and when they

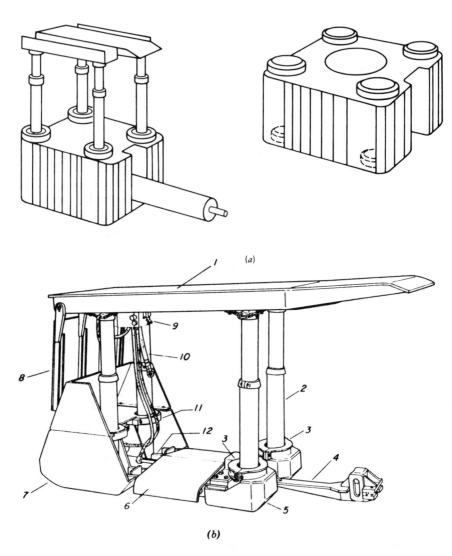

Figure 4.10 Chock type of powered supports (*a*, *b*) [2, 27, 52]. (*b*) 1, Full-width rigid canopy; 2, leg; 3, self-centering leg housing; 4, double-acting ram; 5, front base structures; 6, walkway floor cover; 7, rear base structure; 8, antiflushing shield; 9, hydraulic control valve; 10, hydraulic hoses; 11, stabilizer; 12, frame bars.

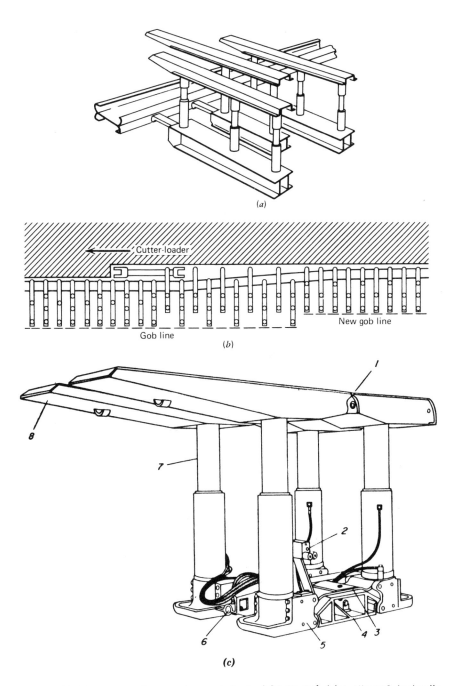

(a)

Cutter-loader

Gob line

New gob line

(b)

(c)

Figure 4.11 Frame type of powered supports (a, b, c) [2, 29, 52]. (c) 1, Hinge; 2, hydraulic control assembly; 3, leaf-spring thruster; 4, center base; 5, footplates with centering brace; 6, shifting cylinder; 7, leg; 8, articulated canopy.

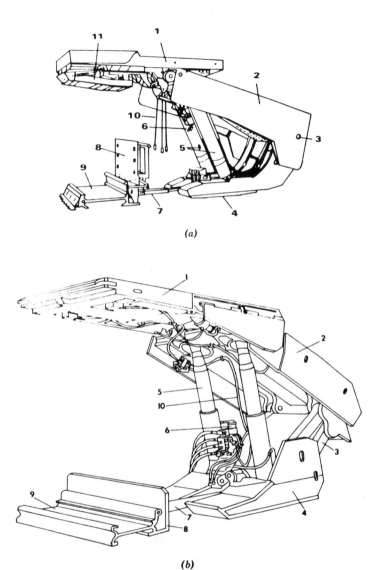

(a)

(b)

Figure 4.12 Shield type of powered supports 29. (a) Caliper shield powered support: 1, canopy; 2, gob shield; 3, hinge; 4, base; 5, legs; 6, hydraulic control valve; 7, hydraulic ram; 8, spillplate; 9, conveyor pan; 10, hose; 11, antispalling plate. (b) Lemniscate shield: 1, canopy; 2, gob shield; 3, lemniscate joint; 4, base; 5, leg; 6, hydraulic control valve; 7, hydraulic ram; 8, spillplate; 9, conveyor pan; 10, hose. (c) Four-leg shield: 1, canopy; 2, gob shield; 3, base; 4, hydraulic ram; 5, spillplate; 6, leg; 7, control valve.

142

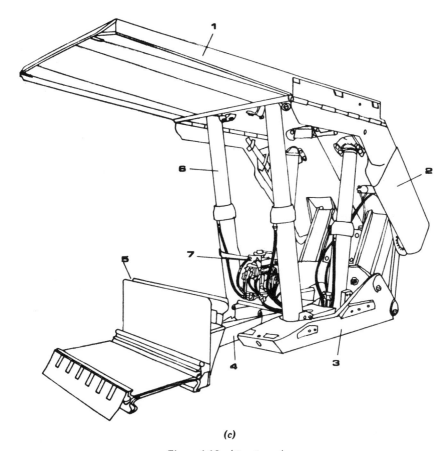

(c)

Figure 4.12 *(Continued)*

shorten the canopy moves forward (Fig. 4.12*a*) [29, p. 236]. In the "limniscate" shield, a special linkage between the base and gob shield maintains a constant distance between the face and canopy tip whether the hydraulic cylinders move up or down (Fig. 4.12*b*). Under such conditions, the system forms a two-armed lever, and the hydraulic cylinders can attack the short lever arm underneath the joint. Recently a four-leg support shield has been developed: the two rear legs react between the base and the shield itself, while the two forward legs react between the base and roof canopy (Fig. 4.12*c*) [29].

4.4.3   Description of Powered Supports

All powered supports, regardless of type, consist of a canopy, a base, hydraulic legs, and control system. Table 4.3 provides dimensions and the operating data for each type of powered support, giving lower and upper limits.

Canopies.   The canopy size ranges from 1.61 to 9.4 m^2 with maximum roof pressure at yield about 4.2–33.2 kg/cm^2. Frame canopies can be any shape, but rectangular shapes are most popular. A frame canopy covers an area less than 70% of the face. A chock canopy is generally a solid piece, articulated to accommodate steps or cavities in the roof. Solid roof canopies have been found to maintain an average of 41% contact area with the roof; contact area increases to an average of 68% when an articulated canopy is used. Chock canopies range in size from 0.19 to 3.87 m^2 and provide a loading pressure at the yield 7–35 kg/cm^2. The fully supported area using chock supports usually runs between 85 to 90% of the face area. Shield canopies range from 2.6 to 4.2 m^2 in size with an average area of 3.2 m^2 (including gob shield) and an average loading pressure at yield less than 28 kg/cm^2. They protect the entire face area.

Bases.   Bases are available in any size. Generally bases for frame supports are split into two halves, whereas solid bases are used for chock and shield supports. A solid base provides better stability. Each base is provided with skids; the most popular design is a combination of a rear skid and a single split forward skid.

Bases for chock and shield supports contain a center tunnel approximately 25 cm wide, open to the floor over the entire base length so that debris can pass toward the gobs. Guide bars are used to transfer the ram jack force to guide the support units during advance without undue side loads. The bottom side of the front edge is usually leveled over the length of approximately 15 cm to reduce tip load and prevent digging into the floor [29, p. 240].

Floor contact area ranges from 0.8 to 4 m^2 for chock and 0.13 to 2.75 m^2 for frame supports. The optimum size of the base for a specific seam floor is such that its unit loading pressure at yield is less than the bearing capacity of the floor rocks.

Table 4.3. Dimensions and Operating Data for Powered Supports[a]

Type	Yield Capacity (tons)	Dimensions (cm)			Maximum Pressure (kg/cm²)		Clear Working Distance (cm)	Supported Roof (%)	Hydraulic Pressure (kg/cm²)	Canopy Tip Yield Load (t)
		Height	Length	Width	Floor	Roof				
Frame	260	81.3	322.6	91.5	19	0.7	114.3	38	70.2	9.7
	1050	322.5	600	350	91.3	33.2	274.3	70	351	58
Chock	150	71.1	284.5	81.2	15.5	1.5	91.5	85	101.8	8
	800	365.8	269	218.4	65.3	37.6	474.9	90	320	45
Shield	115	61	330	130	4	3.5	160	100	140.5	
	600	460	518	150	77.2	21	287	100	35	
Chock shield	320	94	450	140	16.2	7.4	203.2	100		8
	500	340	482.6	145	28	21	254	100		14.3

[a]See reference 29.

145

Legs (Jacks). The bore diameter of the hydraulic legs of powered supports ranges from 10 to 30 cm with operating pressures of the hydraulic pump. When the legs are raised against the roof, the total load exerted on the roof is

$$P = P_i \cdot A \cdot n \qquad (4.7)$$

where P = total setting load, in kilograms
$\quad P_i$ = operating hydraulic pressure, in kilograms per square centimeter
$\quad A$ = cross-sectional area, in square centimeters
$\quad n$ = number of legs

Thereafter, when the roof starts to cave, the hydraulic legs are forced to retreat, and hydraulic pressure in the legs increases. To prevent the hydraulic piston from dropping all the way down to the bottom of the cylinder and going "solid," a spring-loaded yield valve is provided to each support. When pressure in the legs increases to a certain level, the yield valve will open automatically and release the pressure gradually. The pressure at which the yield valve will open is called the yield pressure; the corresponding load applied on the roof is called the yield load. Most support capacities are designed to incorporate the yield loads [29, p. 241].

Hydraulic Power Supplies. There are four types of hydraulic fluids for powered supports [53, 29, p. 243]: (1) 5% soluble oil-in-water emulsion; (2) 40% water-in-oil emulsion; (3) 50% glycol-in-water solution; (4) refined petroleum-based oil. The basic requirements for powered support hydraulic fluids are low cost, low viscosity, nonflammability, and high resistance to chemical change upon contact with air. Furthermore, the fluid should be highly resistant to foaming, because entrapped air not only generates heat when fluid is compressed but also causes mechanical damages when air bubbles collapse under loading. Lubricity and corrosion protection is also important to protect moving parts.

Control Systems. Supports can be controlled in various ways: (1) individual support manually; (2) individual support manually from the neighboring unit; (3) manually from selected points at the face; (4) automatic control from the gateway. The first method of control, used in early designs, has safety difficulties. The second method is used quite extensively. The third method is becoming more popular and is safer. The fourth method is adopted only with a fully automatic operation of the cutting machine as well.

4.5   DESIGN OF POWERED SUPPORTS

Geological and stress conditions are of the utmost importance in the design of powered supports. These factors can affect the strata control as well as the cost item of the support system. For example, if low-yield-capacity supports are used in a strong roof, hardly caved, the pressures will not be met by the supports, the roof will not cave as the face advances, and excessive pressures will result in heavy upkeep expenses on supports. Conversely, if a high-yield support is used in a weak roof, there will be intrusions to the roof, and the unnecessary use of high-cost powered support increases the expense. Thus, the correct capacity of powered supports should be chosen to meet the conditions.

In designing powered supports there is no one established set of formulas or systems. Almost every country has established its own systems. Thus we describe the systems by country.

4.5.1   Dimensions Related to Supporting

Yielding Pressure.   There is a relation between the yielding and setting or operating pressures as follows:

$$P_y = 1.25 P_i \tag{4.8}$$

where P_y = yielding pressure, in kilograms per square centimeter
       P_i = operating pressure, in kilograms per square centimeter

Distance between Supports.   This distance is important in calculations and depends upon roof and floor conditions, bearing capacity of the support, gob conditions (cavings), and rate of advance. It is usually taken as 1.2 m from center to center. It is marked as c in Fig. 4.2.

Unsupported Face Distance. There is always a small distance between the coal at the face and the end of canopy. This distance increases as the winning machine cuts. It may change from 0.25 to 0.8 m according to the depth of cut. It is shown as l_0 in Fig. 4.2.

Load Density. Load density is given by the formula

$$n = \frac{F}{(l_s + l_0)c} \tag{4.9}$$

where n = load density, in tonnes per square meter
  F = carrying capacity of the support, in tonnes
  l_s = length of canopy, in meters
  l_0 = length of unsupported face, in meters
  c = distance between supports, in meters

Maximum and Minimum Heights. "Maximum" and "minimum" define the working heights of the supports according to the geological conditions and convergence evaluation of the face. Owing to changes of the seam thickness, some coal is left at the roof as shown in Fig. 4.13. The working heights are given by the following expression [2, pp. 595, 597]:

$$\log \frac{h_{max}}{1.1 h_{min}} = 1.704 \frac{m'}{m_{av}} \tag{4.10}$$

$$h_{min} = m_{av} - m' - c \cdot l \tag{4.11}$$

where h_{max} = maximum height, in meters
  h_{min} = minimum height, in meters
  m_{av} = average thickness, in meters
  m' = geological deviations in thickness, in meters
  c = average convergence, in millimeters per meter
  l = width (supported span) of the face, in meters

As a numerical example, let us calculate different heights in an average seam of m_{av} = 2.5, of deviations m' = 0.375 m.

$$\log \frac{h_{max}}{1.1 h_{min}} = 1.704 \frac{0.375}{2.5} = 0.2556$$

$$\frac{h_{max}}{h_{min}} = 1.1 \times 1.8014 = 1.98$$

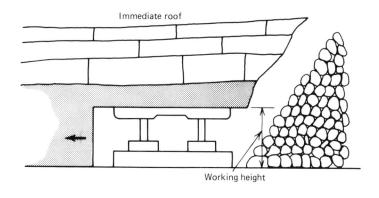

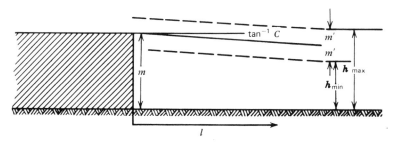

Figure 4.13 Working seam heights for powered supports [2].

This means that the support should have a height almost twice that of the lowest working condition. In Eq. (4.11) numbers taken from Table 4.4 can be used to calculate minimum thicknesses for various face widths.

4.5.2 German System

The carrying capacities of chock and shield powered supports are calculated as shown in Fig. 4.14 [55, 2, p. 605].

For chock supports (Fig. 4.14a), using a safety factor of n, with $K = 1.5$, $\gamma = 2.5$ t/m^3,

$$F_{max} = 5nm \tag{4.12}$$

where F_{max} = maximum carrying capacity of chock support, in tonnes per square meter

m = seam thickness, in meters

n = factor of safety, in general taken 2

Table 4.4 Recommended Minimum Heights for Powered Supports for
Different Seam Thicknesses[a]

Average Seam Thickness (m)	Convergence (mm/m)	Geological Deviations (m′)	Minimum Powered Support Heights		
			$l = 1.75$ m	$l = 2.5$ m	$l = 3.0$ m
0.70			0.68	0.65	0.63
0.80	40	0.05	0.78	0.75	0.73
0.90			0.88	0.85	0.83
1.00		0.05	0.96	0.93	0.90
1.50	50	0.10	1.51	1.48	1.45
1.80		0.15	1.86	1.83	1.80
2.00		0.15	2.06	2.03	2.00
2.20		0.20	2.26	2.20	2.16
2.40		0.20	2.46	2.40	2.36
2.60	80	0.20	2.66	2.60	2.56
3.00		0.20	3.06	3.00	2.96
3.20		0.25	3.26	3.20	3.16

[a] See references 2 and 53.

For shield supports, as seen in Fig. 4.14b and c,

$$F = \frac{L_r}{L_f} R \qquad (4.13)$$

where F = carrying capacity, in tonnes
  R = piston reaction, in tonnes
  L_f = distance of carrying load to the back hinge, in meters
  L_r = distance of piston to the back hinge, in meters

4.5.3 English System

In the English system the weight of the immediate roof is taken into
consideration, as shown in Fig. 4.14, and is given by the following:

$$F_{min} = \gamma \frac{m}{K - 1} \qquad (4.14)$$

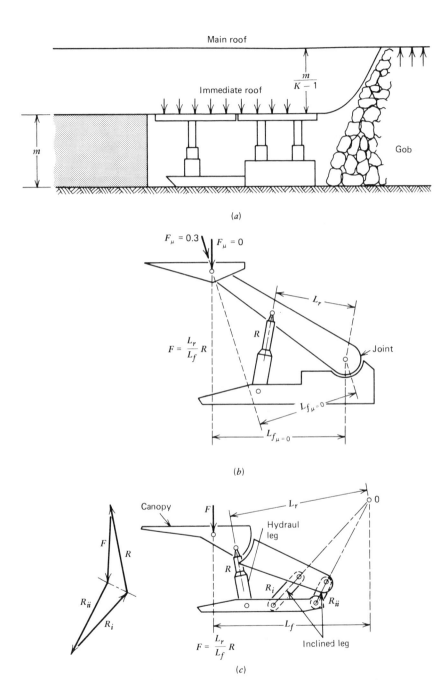

Figure 4.14 Loads on chock and shield powered supports [2, 55].

where F_{min} = minimum capacity of support, in tonnes per square meter

m = thickness of the seam

γ = density of immediate roof, in tonnes per cubic meter

K = factor of expansion of the immediate roof; may be taken as 1.5 for average

To obtain the values of K, refer to Section 1.3.3 "Pressure in Longwalls."

W. Wilson [56, p. 57] has analyzed loads under level and inclined conditions. In inclined seams the load on the supports is given by the following formula, as shown in Fig. 4.15, and various loads calcu-

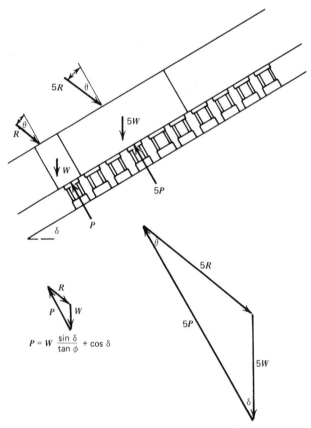

Figure 4.15 Loads in inclined seams [56].

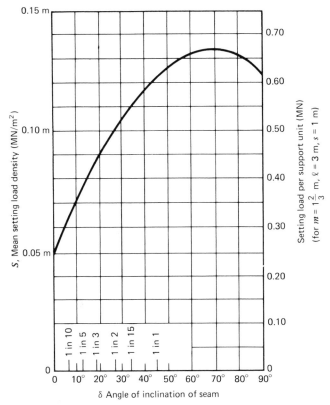

Figure 4.16 Variation of loads with the angle of inclination [56].

lated are plotted against the angle of inclination given in Fig. 4.16 [56, p. 585–586].

$$P = W \left(\frac{\sin \delta}{\tan \varphi} + \cos \delta \right) \tag{4.15}$$

where P = fixing load of support, in tonnes

W = weight of the block on the support, in tonnes

δ = angle of inclination, degrees

φ = angle of friction between immediate and main roof

4.5.4 Austrian Systems

According to Sigott [58], the bearing capacities of powered supports can be calculated as illustrated by Fig. 4.17. The main thought is that

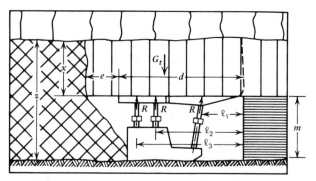

Figure 4.17 Load on the powered supports [2, 58].

the immediate roof is supported by the hydraulic systems of the powered chocks or frames [2, p. 617]. The moments of hydraulic supports should be greater than the moment of dead weight of the immediate roof.

$$R(l_1 + l_2 + l_3) \geqslant G_t \frac{d+e}{2} \qquad (4.16)$$

$$G_t = \beta(d+e)t\gamma = \beta(d+e)\frac{m}{K-1}\gamma \qquad (4.17)$$

$$R = \frac{nR_0}{3} \qquad (4.18)$$

$$l_1 + l_2 + l_3 \cong 2d \qquad (4.19)$$

$$\frac{nR_0}{3} 2d \geqslant \beta \frac{(d+e)^2}{2} \frac{m}{K-1}\gamma \qquad (4.20)$$

$$R_0 \geqslant \frac{3}{4} \frac{\beta}{n} \frac{(d+e)^2}{K-1} \frac{m}{d}\gamma \qquad (4.21)$$

where R_0 = minimum bearing capacity of one hydraulic unit, in tonnes

β = diminishing factor, usually taken as 0.9

n = number of units of power support frame (or chock) per linear meter of face

d = the length of canopy or distance between the back of support and face line (Fig. 4.17), in meters

e = the distance between back of support and uncaved roof (Fig. 4.17), m

m = seam thickness, in meters

K = expansion factor, usually taken as 1.4–1.6

γ = density of immediate roof, in tonnes per cubic meter

According to Eq. (4.21), as e (uncaved area) is increased, the support should be greater (stronger); to diminish this area, stronger supports should be chosen.

4.5.5   French System

The French system bases calculations for powered supports on convergence measurements at the face. According to Josien-Gouilloux [59] the following formula and curves drawn accordingly, are utilized in French coal mines:

$$CvT = (qW)^{3/4}H^{-1/4}\left(\frac{6800}{P_a} + 66\right) \qquad (4.22)$$

where CvT = convergence at the face, in millimeters per meter of advance

W = thickness of the seam, in meters (between 0.8 and 3 m)

q = subsidence factor: 1 for caving; 0.6 pneumatic stowing; 0.15 hydraulic stowing

H = depth below surface, in meters (between 100 and 1000 m)

P_a = load carrying capacity of support, in tonnes per meter of face length).

The application of Eq. (4.22) is given in Fig. 4.18 [59, p. 53] for a depth of 500 m for different equivalent thicknesses qW. The convergence diminishes rapidly for 10–100 t/m bearing capacities, and convergence above 40 mm/m should not be allowed.

Roof conditions categorized A, B, and C are shown for the desired load carrying capacities (in tons per meter) against equivalent seam thicknesses qW in Fig. 4.19 [59, p. 54]. The roof condition A refers to a "thick strong roof" (over 1 m) made of sandstone, sandstone-shale, conglomerate, strong limestone. There are few fractures, and the roof is cut into large blocks by parallel fissures sloping toward

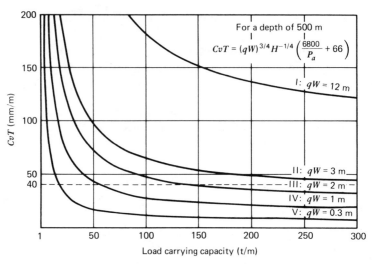

Figure 4.18 Relationship between average convergence and the load carrying capacity per linear meter of the support [59].

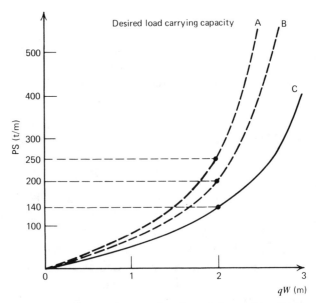

Figure 4.19 Desired carrying capacity as a function of the working thickness at 500 m depth [59]. Equivalent working thickness qW in meters. A, thick, strong roof; B, stratified, strong roof; C, fragile roof.

the goaf. B refers to "stratified strong goaf" consisting of thinner strata, cracked in a finer mesh than A. The curve C refers to a "fragile roof" (coal, marly schist, shale) which comes off in small blocks as soon as they are exposed at the face. As shown in the figure, 140 t/m support is sufficient for the fragile roof, whereas at least 250 t/m capacity is needed for the thick strong roofs [59].

4.5.6 Polish System

The Polish system of calculating powered supports is based upon the openings at the face. The average bearing capacity of an area supported by three units is shown in Fig. 4.20 [60, 2].

$$P_0 = \frac{P_1 + P_2 + P_3}{F} n \qquad (4.23)$$

where P_0 = average carrying capacity, in tonnes per square meter
P_1 = nominal load of one unit, in tonnes
P_2 = load on the unit when advancing, in tonnes, taken as zero
P_3 = carrying load of the unit just set, in tonnes
F = the area of the face covered by three supports, in square meters
n = efficiency factor of supports, taken around 0.8

As a numerical application let us take an area covered by three supports placed at 140-cm intervals: The width of the face, as shown in Fig. 4.20, is 0.60 m unsupported at the back + 3.65 m supported by the canopies + 0.35 m unsupported at the front, all totaling 4.6 m. The nominal carrying load is P_1 = 70 t per leg, and the newly set support carrying a load is P_3 = 23 t per leg. Each support is equipped by four legs. Then

$$P_1 = 4 \times 70 = 280 \text{ t}$$

$$P_2 = 0$$

$$P_3 = 4 \times 23 = 92 \text{ t}$$

$$F = 4.60 \times (3 \times 1.4) = 19.32 \text{ m}^2$$

$$P_0 = \frac{280 + 0 + 92}{19.32} \times 0.8 = 15.4 \text{ t/m}^2$$

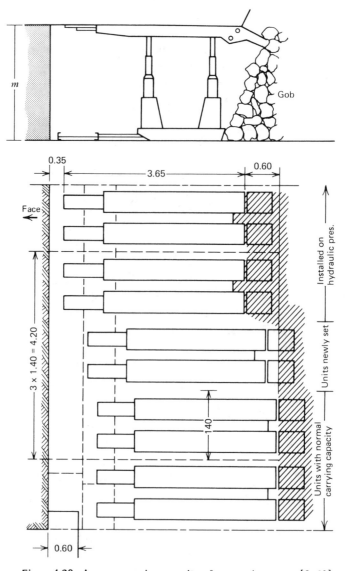

Figure 4.20 Average carrying capacity of powered supports [2, 60].

Table 4.5   Roof Indices for Powered Support Assessment[a]

Category	Roof Index L	Characteristics of the Roof
I	$0 < L < 18$	Roof caves as soon as exposed Coal is left to hold the roof
II	$18 < L < 35$	Roof caves easily as the support is drawn
III	$35 < L < 60$	In smaller L values, roof caves In larger values roof gets strong but caves easily
IV	$60 < L < 130$	Roof usually caves but caving becomes more difficult as L becomes bigger
V(a)	$130 < L < 250$	Roof is strong, it does not cave It caves with special attention (blasting, etc.)
V(b)	$L > 250$	Roof is very strong; it does not cave; stowing systems of mining should be applied

[a]See references 2 and 60.

To evaluate proper support, the roof conditions are assessed. In the assessment roofs are divided into five groups with "roof indices" as described in Table 4.5 [60, 2].

The roof index L can be taken from Table 4.5, but it can be calculated by the following, as well [60]:

$$L = 0.0064 \sigma_b^{1.7} K_1 K_2 K_3 \qquad (4.24)$$

where L = roof index

σ_b = uniaxial compressive strength of roof rock measured on dry specimens in the laboratory, in kilograms per square centimeter

K_1 = in situ strength coefficient, 0.33 for sandstone, 0.50 for siltstone

K_2 = fatigue coefficient, 0.70 for sandstone, 0.60 for siltstone

K_3 = in situ water content coefficient, 0.60 for sandstone

In a numerical application calculated for a sandstone of compressive strength 500 kg/cm², the roof index becomes

$$L = 0.0064(500)^{1.7} \times 0.33 \times 0.70 \times 0.60$$

$$\cong 34$$

which falls to category II in Table 4.5.

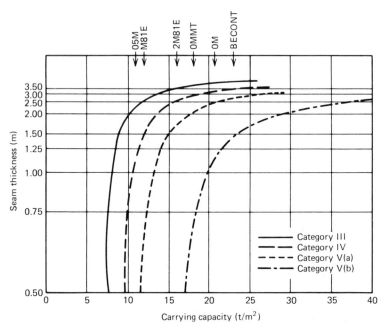

Figure 4.21 Supporting capacities of powered supports under several roof conditions [2, 60].

The carrying capacities of powered supports in tonnes per square meter are given according to seam thickness and roof conditions. Several roof indices (for categories III to V) are plotted in Fig. 4.21 [60, 2]. Some firms that produce powered supports are indicated in the figure.

The unsupported distance in front of the face is related to the roof index and the "slabs" formed at the roof, as shown in Fig. 4.22 [60, 2]. The height of slabs h is a criterion of the working conditions if:

$h < 10$ cm, normal working conditions
$20 < h < 10$ cm, difficult conditions
$h > 20$ cm, dangerous conditions

The normal and difficult conditions are shown as two curves in Fig. 4.22. The allowable unsupported areas for different roof conditions are given in Table 4.6 [60, 2].

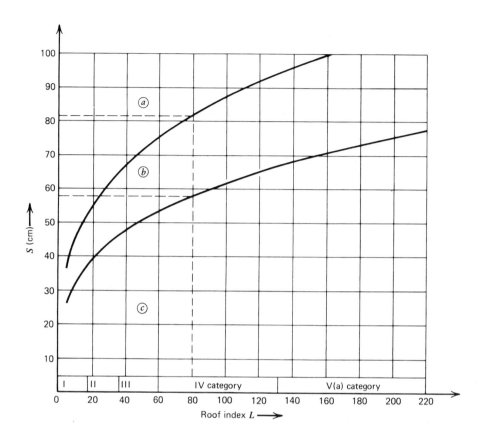

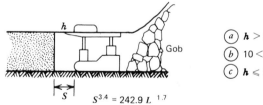

$$S^{3.4} = 242.9 \, L^{1.7}$$

ⓐ	$h > 20$ cm dangerous
ⓑ	$10 < h \leqslant 20$ cm difficult normal
ⓒ	$h \leqslant 10$ cm normal

Figure 4.22 Unsupported distance according to the roof indices [2, 60]. $h_b > 20$ cm, dangerous; $10 < h < 20$, difficult; $h < 10$ cm, normal.

161

Table 4.6   Allowable Unsupported Areas of Different Roof Conditions[a]

Category	Roof Index L	Roof Conditions	Allowable Area (m^2)
I	0–18	Tabulated siltstone, water bearing, cracks unsupported	1
II	18–35	Broken siltstone, water bearing, breakable roof	1–2
III	35–60	Siltstone or mudstone, easily caved	2–5
IV	60–130	Typical caving roof, strong siltstone, mudstone, coarse grained sandstone	5–8
V(a)	130–250	Strong siltstone, fine-medium grained sandstone (strong roof)	8
V(b)	>250	Uncaved roof-stowing systems	

[a]See references 2 and 60.

4.5.7   American System

In this American method, developed by U.S. Bureau of Mines, it is assumed that the immediate roof behaves like a cantilever beam [52]. It breaks in front of the face at a distance equal to seam height, and the roof to be supported extends from the end of the overhang to the assumed break. Three cases are shown in Fig. 4.23 [52, 29, p. 246].

In the first case, there is a minimum gap between the tip of the support and the coal faceline; the second case is similar to the first except that a cut has been made in the face; in the third case there is a large overhang. The static load to be supported by the supports for all three cases can be generalized as follows:

$$W = L_i SwH \qquad (4.25)$$

$$T = W \qquad (4.26)$$

where W = weight of the immediate roof to be supported
L_i = length of the beam

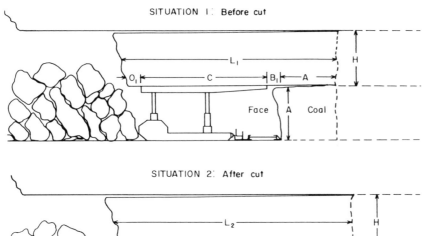

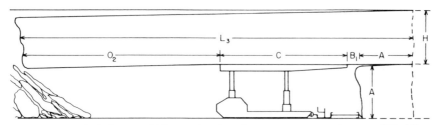

Figure 4.23 U.S. Bureau of Mines estimation of power support requirements [27, 52].

i = case numbers: 1, 2, 3
S = average spacing between the supports
w = average weight density of the roof rock
H = thickness of the immediate roof
T = minimum rated yield load of the support

The weight of the immediate roof calculated in accordance with Eq. (4.24) for the longwall chock and longwall shield supported faces is shown in Fig. 4.24 [29, p. 247]. In the calculation the immediate

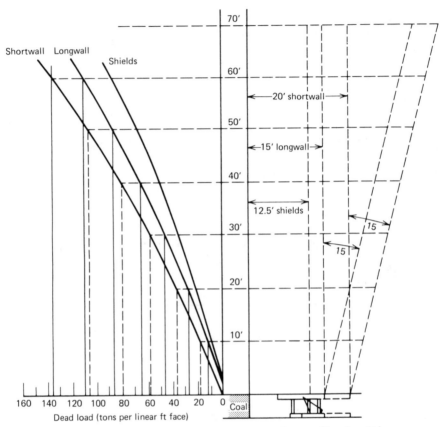

Figure 4.24 Method of determining support resistances [Joy Co., 29].

roof does not overhang the gobs; it caves at the gob edge of the support at a caving angle of 15° toward the gob, the most frequently observed caving angle in the Eastern United States coalfields. As an
example of using the figure, assume a case where the immediate roof
is 20 ft thick. First locate the 20-ft horizon in the roof. Extend this
line horizontally to the left until it intersects the curve lines marked
"shortwall," "longwall," and "shield." Drop a vertical line from each
intersection until it hits the horizontal axis of "dead load." The
points of intersection indicate that the dead load per linear foot of
face is 23 tons for the shield, 28 tons for the longwall chocks.

4.6 ADVANTAGES AND DISADVANTAGES OF POWERED SUPPORTS

4.6.1 Advantages of Powered Supports

Low Convergence. Hydraulic systems control the roof very efficiently. Large canopies hold the roof effectively. The convergences measured in powered and conventional (hydraulic prop plus articulated caps) are shown in Fig. 4.25 [61, 2, p. 643].

High Production. Owing to mechanization, the systems are able to advance as much as 5-6 m a day. This increases the production (more than 1500-2000 tons), decreases the number of faces required, and obtains concentration in the mine workings.

Safe Production. Effective roof control has minimized accidents caused by roof falls. Accident rates in the British collieries are shown

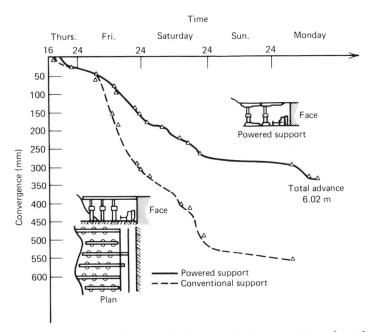

Figure 4.25 Convergence in powered and conventional supported faces [2, 61].

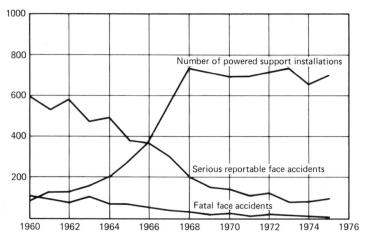

Figure 4.26 National Coal Board powered support installations and accident rates [62].

in Fig. 4.26 [62, p. 678]. Serious and fatal accidents at the face declined greatly with the increase in powered support installations between 1960 and 1976.

High Efficiency. The output per conventional manshift has increased tremendously compared to the output of conventional supporting systems. The efficiencies, labor costs, and other related statistical data are given in Table 4.7 [63, 2, p. 645].

4.6.2   Disadvantages of Powered Supports

Capital Cost. Powered supports require high capital expenditure. Unless there are large panels available, they may not be justified.

High Cost of Upkeep. The cost for upkeep is much higher than the cost for conventional supports.

Qualified Labor. Powered support systems do require highly qualified labor.

Geological Specifications. Geological specifications are difficult to meet. Large panels, small fluctuations in seam thickness, and conditions of mechanical workability should be met.

Table 4.7   Comparison of Powered and Classical Supporting System[a]

Rate	Advance-ment (m/day)	Face Product (t/day)	Area Opened (m²/t)	Support Efficiency		Labor Cost	
				Area (m²/shift)	Area (m²/t)	(DM/m²)	(DM/t)
Classicial Supports: Friction + Hydraulic Props							
Minimum	0.73	516	207	12.3	6.6	16.56	6.65
Maximum	3.78	966	490	28.0	11.9	8.78	3.61
Average	2.89	728	356	19.7	9.8	11.37	5.48
Powered Supports							
Minimum	1.61	735	303	28.3	9.3	11.08	4.56
Maximum	5.52	1766	732	71.2	27.5	3.88	1.79
Average	4.35	1173	552	49.9	16.2	6.77	2.98

[a] See references 2 and 63.

4.7   APPLICABILITY OF POWERED SUPPORTS

Powered supports, although very effective in supporting and very valuable in mechanization, have both geological and technical limitations in application.

4.7.1   Roof Conditions

The roof should cave. If it does not cave or hangs and caves suddenly, it is not suitable for longwall mining. Some stowing system should be used. The most suitable root caves as the support advances. However, when a very weak roof crumbles rather than holds, a part of the coal is left to support it.

4.7.2   Floor Conditions

The floor should be strong enough to resist "intrusions." Intrusion of soft floors is troublesome for advancing and also makes the roof conditions difficult owing to high convergence. Some coal may be left if the coal is hard.

4.7.3   Seam Thickness

The thickness of the seam and its regularity is important. Supports can be lengthened and further increased by some additions. However, great irregularities cannot be met. Cutting low sections may be difficult for the winning machines and washing facilities. Therefore, a good survey of the seam should be made before choosing the proper supports. The maximum thickness that can be worked, at present, is 5 m. Thicker seams should be worked out in slices or by recovering the caved coal from the back by special arrangements.

4.7.4   Seam Inclination

Although the best operation is on level seams up to 8° of inclination, by special additions, seams up to 35° of inclination can be worked out by powered supports.

4.7.5   Small Faults

Small faults can be passed by cutting the roof or floor rock. The winning machines should be able to cut these areas. Too many faults are troublesome; they slow down the advancement and are difficult for washing facilities. Large faults are impossible to cut. New development work is needed. The best panels are large, with few or no faults, so that once the face is set, a long undisturbed operation is fulfilled.

4.7.6   Water at the Face

Water at the face is detrimental and corrosive to the supports. Under these conditions either the panel should be drained by drilling, or special anticorrosive supports should be chosen. Water is always a handicap, and miners hate to work under the wet conditions.

4.7.7   Life of the Panel

The width of the panel should be large enough to warrant powered supports. It takes 15-20 days to install the equipment, which adds to the cost of the coal; the expense is least in large panels. However, very large panels may have expenses in the upkeep of the gateways, although this is not difficult to solve. The optimum width is found to be 800–1000 m.

4.7.8 Face Length Rate of Advance

Production depends upon the face length and the rate of advance. These are important factors on the life of the panel and number of working faces. The haulage capacity should be chosen to accommodate these factors. The practical rate of powered supported faces is about 5–6 m per day, as average.

4.7.9 Number of Shifts

For the highest production, the faces should operate continuously, but this is not possible all the time. Two shifts per day have been found to be quite practical, leaving one shift for upkeep and preparations. However, the average is moving toward 2.5 shifts a day as the number of mechanized faces increase [64].

4.7.10 Hints for Good Installation

To achieve regular production with the fewest disturbances, the following hints should be observed:

1. The face should be straight.
2. The gate roads should be large enough to carry production and service the face. The deformed places should be enlarged right away.
3. The location for the erection of support at the face should be ready in advance.
4. The transport of the supports should be planned and delays eliminated.
5. The supports should first be installed at the surface, checked for everything, and adjustments made, if needed.

6. It is easier to install the conveyor first and then install the supports according to the conveyor.

7. Hoses should be connected and special attention should be given to the connections and leakage.

8. A good signaling system is required at the gateways. Telephones are quite necessary.

9. The hydraulic pump system is essential. Special care should be given to this system. The lower gate is the preferred place for installation.

10. An oil–water mixture as a fluid is used, and the best proportion should be maintained throughout the operation.

11. Finally, during the operation convergence should be measured. The strata movements should be observed. Excessive weights on the supports should be checked and extra supports added, if necessary.

CHAPTER 5

Concrete Supports

5.1 IMPORTANCE OF CONCRETE

The use of concrete as a support material in mines is limited. However, its use is becoming more frequent and is considered a necessity in the following:

Shafts

Large section galleries like pit-bottom arrangements, pump-stations, and tippler stations

Lining to eliminate spalling

Water sumps

Dams for water, fire, explosion

Artificial roof for multilayer mining

As a material concrete is designed mainly to withstand compressional stresses. However, by including steel bars and forming a new material known as "reinforced concrete," tension stresses can be met. Both material and its application form an essential part of civil engineering. This chapter touches only the high points of the study of concrete, especially as related to the applications just listed.

5.1.1 Advantages of Concrete

The advantages of concrete as a supporting material over timber and steel can be summarized as follows:

1. As a compressive material it has high strength and is quite economical.
2. The constitutents of concrete (cement, aggregates, water) are easily obtained in any quantity.
3. The characteristics of these constituents are straightforward.
4. Concrete can be easily installed in most places.
5. Application (mixing, transporting, pouring) can be mechanized and the cost reduced.
6. It is the safest material in respect to fire resistance.
7. Because it gives a smooth surface to linings the resistance to the flow of air is diminished.
8. It is not affected by atmospheric conditions and has a long life.

5.1.2   Disadvantages of Concrete

Some disadvantages to the use of concrete, which should be evaluated carefully before application, should be summarized as follows:

1. It has very low tensile strength, so it should not be designed for meeting conditions of tension, or, if used under these conditions, should be reinforced by steel.
2. It breaks suddenly without giving any warning like fibering with timber or deformations in steel.
3. Broken concrete has no value. Unlike steel or timber, it cannot be reused, and should be removed.
4. Since the compressive strength of concrete is largely influenced by the processing, the amount of constituents, curing time and so forth, should be controlled carefully. Its use requires more supervision than does the use of other materials.

5.2   CONSTITUENTS OF CONCRETE

Concrete is essentially a mixture of cement, aggregates (gravel or broken stone, sand), and water. It is combined using different ratios depending on its intended use. Some materials may be added to meet requirements or to shorten the setting time.

5.2.1   Cement

Cement is the most important constituent of concrete. When mixed with water it forms a hard material that holds together the added material (aggregates). "Portland cement" is most frequently used in mining operations. However, in special cases, quick-setting cements are used. Every nation has set standards on Portland cement, and it is manufactured under these specifications and should be used according to these recommendations. The list of British and American standards are added at the end of the chapter as appendices.

Cement sets after mixing with water. The minimum time for setting is one hour and maximum, 10 hours. Temperature is an important factor; setting is shortened in warm conditions.

The "dosage," the amount of cement expressed in kilograms in a mixture of 1 m^3 volume, is given by the following:

$$M_c = \frac{550}{\sqrt[5]{D_{max}}} \tag{5.1}$$

where M_c = minimum dosage, in kilograms of cement per cubic meter of concrete

D_{max} = the largest aggregate size, in milimeters

It can be seen that, as the aggregate size increases, the amount of cement decreases, making the concrete more economical. It is advantageous to use large aggregates wherever possible. In general, the dosage is 300–350 kg/m^3. In special cases where higher strength is required, dosages of 400–450 kg/m^3 are used.

5.2.2   Aggregates

"Aggregate" is a mixture of sand and gravel or broken rock. Sand is the portion 0–7 mm in size, and gravel is 7–30 mm. Sizes larger than 30 mm are not used where transportation by pipeline is desired. Sand and gravel form the skeleton of concrete and minimize shrinkage in volume during setting.

There should be a "granulometric curve" to show the percentages of different sizes. The ideal granulometric distribution is shown in Fig. 5.1a for gravel and Fig. 5.1b for sand [65, 2, p. 658]. Granulomery of the aggregates should be around these curves.

5.2.3   Other Constitutents

Additional materials are included in concrete to change the setting time and fluidity.

Calcium Chloride (CaCl₂).   Calcium chloride is the most frequently used agent for decreasing the curing time of concrete. The concentration is 2%, added to the mixing water. This decreases the curing time to 1–3 days. Warm temperature also shortens the setting time.

Sugar.   Sugar is an important element included to retard setting. It can prolong it in proportion to the concentration.

Fly Ash.   Fly ash is obtained from power plants and is an important addition to concrete that increases its fluidity in pipeline transportation. Pure silica (SiO_2) and bentonite in fine aggregate (0–0.2 mm) are also materials used to aid the fluidity. In concrete with high dosages of cement ($\geqslant 350$ kg/m³), such additions may not be required. However, in poor concretes (200–250 kg/m³) fine aggregates should make up 10% of the total aggregate weight.

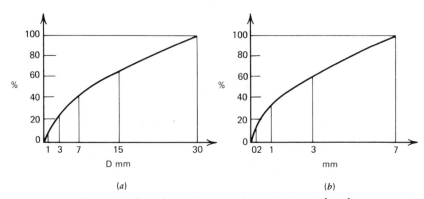

(a)   (b)

Figure 5.1   Granulometric curves of gravel and sands [2, 6].

Table 5.1 Coefficients of Working Conditions of Concrete[a]

	Coefficient A	
Working Conditions	Gravel	Broken Rock
Wet (2–6 cm sinking)	45	50
Plastic (7–12 cm sinking)	50	63
Fluid (>12 cm sinking)	58	74

[a] See reference 2.

5.2.4 Water

Water is an important factor in concrete, making the hydration of the cement and forming the "fluidity" of the mixture. The amount of water needed depends upon the granulometry of the aggregates and the compressive strength desired. It is given by the following formula:

$$M_w = A(7 - K) \qquad (5.2)$$

where M_w = amount of water, in kilograms per cubic meter
   A = a coefficient according to the working conditions (Table 5.1)
   K = fineness modulus,* comulative percentage of aggregate larger than a given sieve opening size

It can be seen that, as the concrete gets more fluid, more water is needed. Broken rock aggregates require more water than do the gravels.

5.3 ENGINEERING CHARACTERISTICS OF CONCRETE

5.3.1 Water-Cement Ratio

The water-cement ratio is the most important factor in the compressive strength of concrete. It is also an important factor in the transportation of concrete in pipes. The effect of α (water-cement ratio

*The fineness is defined as the sum of the cumulative percentages retained on the standard seives divided by 100.

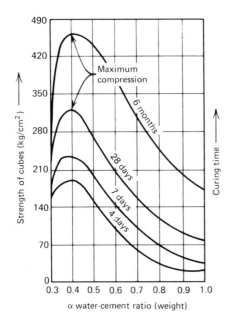

Figure 5.2 Compressive strength of concrete in terms of α (water-cement) ratios [2]. Normal Portland cement mixing ratios: cement 1 : sand 2 : gravel 4.

in weight) is shown in Fig. 5.2. It can be seen that in all curing time the strength reaches a maximum at $\alpha = 0.4$ and decreases afterward with increasing α values [2, p. 661].

Many formulas are given to express the compressive strengths in terms of α ratios. According to Abrams [66]

$$\sigma_b = \frac{A}{B^\alpha} \tag{5.3}$$

According to Bolomey [66]

$$\sigma_b = K\left(\frac{1}{\alpha} - 0.5\right) \tag{5.4}$$

According to Graf [67]

$$\sigma_b = \frac{K_n}{a} \cdot \frac{1}{\alpha^2} \tag{5.5}$$

where σ_b = compressive strength, in kilograms per square centimeter after a certain curing time

α = water-cement ratio in weight

A = coefficient for 28 days of curing, 950

    B = coefficient for 28 days of curing, 9
    K = coefficient for 28 days of curing, 180
       = coefficient for 7 days of curing, 150
    K_n = compressive strength of cement (according to the many standards 400 kg/cm²)
    a = coefficient of workmanship, good: 4, medium: 6, poor: 8

The amount of water is also important for the working condition of the concrete. Various amounts of water are given in Table 5.2 [2, p. 662].

5.3.2 Compaction

The "compaction" of concrete is the volumetric sum of the solid materials (cement plus aggregates) in 1 m³ of concrete. It is the reverse of the "porosity." It is well known that the compressive strength of the concrete decreases with the porosity, as given by the Feret formula [66]

$$\sigma_b = K \left(\frac{V_c}{1 - \Lambda + V_c} \right)^2 \tag{5.6}$$

$$p = 1 - \Lambda \tag{5.7}$$

where σ_b = compressive strength of concrete after a known curing time
    K = coefficient, changing according to curing time and granulometry of the aggregates

Table 5.2 Amount of Water for Different Working Conditions[a]

Condition	α (Water-Cement)	Water (kg/m³)	Strength (Cubes, 28 days, kg/cm²)
Very dry	0.52	130	270
Wet	0.58	145	232
Plastic	0.64	160	201
Fluid	0.70	175	177
Liquid	0.76	190	156

[a] See reference 2.

V_c = the volume of cement in 1 m^3 of concrete
Λ = compaction
p = porosity

To decrease porosity, that is, increase the compaction, vibrating hammers or similar tools are used.

5.3.3 Granulometry of Aggregates

The size and shape of aggregates are important factors in the workability and compressive strength of concrete. The amount of water in a mixture is a function of the granulometry. As strength depends on the water-cement ratio, the maximum size of coarse aggregate increases the strength and water decreases it. The conditions for the aggregates are given in Table 5.3 [68, 2, p. 664].

It can be seen that by using angular-coarse aggregates the ratio of aggregate-cement is diminished and the strength is increased.

5.3.4 Curing Conditions

Curing conditions are of the utmost importance in concrete making. The curing of the cement increases with the time. This is shown in Fig. 5.3 [2, p. 666].

As can be seen in the figure, curing follows a logarithmic curve. The curing may last a year, and in water, a few years. In practical

Table 5.3 Relation between Aggregate Size-Shape and Aggregate-Cement Ratios[a]

Shape of Aggregates	Aggregate-Cement Ratio
Rounded coarse and irregular fine aggregates	6.5
Irregular coarse and irregular fine aggregates	5.5
Angular coarse and irregular fine aggregates	5.2

[a]See references 2 and 68.

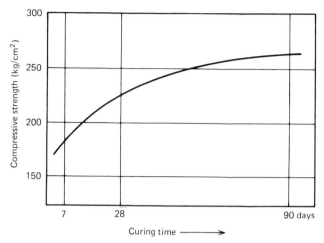

Figure 5.3 Curing time of concrete [2].

work 28 days is taken as the standard for curings and the compressive strength is related to this timing. However, in quick work, a 7-day cure may be taken, which is 70% of 28-day curing. In 90 days the strength increases up to 120% of that for 28-day curing.

The "wet" condition of the concrete is also important. It should be kept in wet conditions for two weeks to obtain complete hydration.

In addition temperature has an effect on hydration. Setting takes place at 15–25°C. The higher temperature shortens the setting time causing shrinkage; the lower temperature retards the curing.

5.3.5 Working Conditions

The following sections describe the three types of concrete used in practice.

Wet Concrete. The α (water-cement) ratio is 0.3–0.5. The cement should stick to the hand if mixed by hand. The "slump" amount is 2–6 cm on Abram's cone.

Plastic Concrete. The α ratio is 0.45–0.65, containing more water. The slump amount is 7–12 cm on Abram's cone.

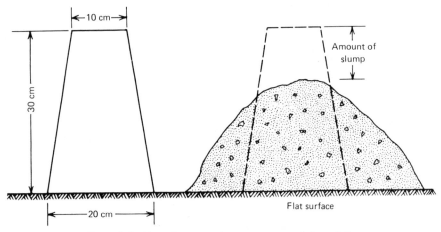

Figure 5.4   Abram's cone and measurements of slump [2].

Fluid Concrete.   The amount of water in fluid concrete is much more than in wet or plastic ($\alpha = 0.6$–1.0) allowing the mixture to be pumped. The slump is 10 cm.

Abram's Cone.   Abram's cone is a measuring system of slump. A circular cone 20 cm at base, 10 cm on top, and 30 cm high is filled in three portions by concrete and compacted by 25 blows of the vibrator in each filling. After 3 minutes, the cone is taken away and the concrete is left alone. The slump (loss of height) is measured as shown in Fig. 5.4 [2, p. 667].

5.3.6   Making of Concrete

For very small jobs, cement is made by hand, mixing all the constituents by shovel. First the cement and aggregates are mixed dry, and a cone is prepared. Then the water is added with continuous mixing.

For larger amounts, a mixer is used. All the constituents (cement, aggregates, water) are added at the same time. They are mixed by the turning action of the mixer for 1.5–2 minutes. For continuous operation, the first ingredient is wet coarse aggregate, then cement, fine aggregates, and water are added and mixed for not less than 1.5 minutes (3–5 minutes are better). After pouring the concrete, the mixer can be cleaned by pressurized water.

The capacity of a mixer is calculated as follows:

$$Q = V \frac{60}{T} a \qquad\qquad (5.8)$$

$$T = \frac{t_1 + t_2 + t_3 + t_4}{60} \qquad\qquad (5.9)$$

where Q = capacity of concrete in compacted form, in cubic meters per hour

a = condition factor, a = 0.65 for wet, a = 0.85 for plastic concrete

T = period of mixing, in minutes

t_1 = charging time, in seconds; usually 25 s

t_2 = mixing time, in seconds; minimum of 90 s

t_3 = discharging time, in seconds; usually 15–20 s

t_4 = idle time, in seconds; that is, the dead time between periods

As a numerical example, let us calculate the concrete produced at a shaft sinking operation by a mixer of 150-liter capacity, at t_1 = 20, t_2 = 120, t_3 = 15, t_4 = 20 s, and a = 0.85 (plastic)

$$T = \frac{20 + 120 + 15 + 20}{60} = 2.9 \text{ min}$$

$$Q = 0.150 \frac{60}{2.9} 0.85 = 2.6 \text{ m}^3/\text{h}$$

5.3.7 Transportation of Concrete

In mining operations concrete is transported by a pipeline. The important factors in pipeline transport are given in the following sections.

Concrete Strength. The mixture is determined by the strength of the concrete desired. For monolithic shaft and gallery lining this is 160–225 kg/cm² in 28 days of curing; in the stowing of gateway ribs the 7-day curing strength should be 150–300 kg/cm².

Granulometry. The maximum size of the aggregate should be 30 mm. For normal concrete, fine-grained aggregate (0–0.25 mm) should be a minimum of 5%. In rich concretes (dosage > 350 kg/m^3) there is no need of fine aggregates.

Slump.   The slump of fresh concrete should be 7.5 ± 2.5 cm.

Pipeline.   At its start the pipeline should have a minimum of 8 m horizontal discharge. It should not have sharp curves and vertical inclinations should be minimized. It should operate without interruptions. In case of stopage, the pump should give strokes every 3–5 min to avoid any setting of the concrete in the pipeline.

5.3.8   Pouring and Maintenance of Concrete

The maximum height in pouring should not be more than 2 m. The thickness of concrete in monolithic lining is about 10–15 cm for wet concrete and about 100 cm for plastic concrete. In adding to older concrete surfaces, the old surface should be brushed to coarsen the surface to improve adherence.

The concrete should be compacted by vibrators. In such an operation a small film of water is produced at the top. The vibration time is about 2–3 min for 1 m^3 of concrete. It should not be prolonged, as it may change the homogenity of the concrete. The vibration should be completed after every 40 cm of concrete thickness. The vibration should also be used on the surface of steel forms. The surface of the concrete should be kept wet for two weeks.

The times for removing the forms used when pouring concrete are given in Table 5.4 [2, p. 671].

Table 5.4   Time of Removing of Forms (Days)[a]

Cement	Side Forms	Floor Forms	Columns, Large-Span Floors
Normal Portland	4	10	28
Quick Portland	3	5	10

[a] See reference 2.

5.3.9 Strength of Concrete

Concrete is a well-studied material, and specifications are drawn in every phase of its use. Lists of British and American standards are given at the end of the chapter as appendices [66]. All the information can be found in these standards.

To give examples, the three most widely used concrete strengths in Turkish standards are given in Table 5.5 [65, 2, p. 672]. The British standards for two concretes are given in Table 5.6 [66, p. 291]. The 28 compressive strengths of some American standards are given in Table 5.7 [69, p. 34].

Table 5.5 Compressive Strength of Concretes (Turkish Standards 500)[a]

Type of Concrete	Cubes 20 X 20 X 20 cm (kg/cm^2)	Cylinders 15 cm ϕ X 20 cm (kg/cm^2)	Ratio (cube/cylinder)
B 160	160	140	1.14
B 225	225	195	1.15
B 300	300	240	1.25

[a] See references 2 and 65.

Table 5.6 Strength of Concretes (British Standards 12)[a]

	1 day	7 days	28 days	365 days
Standard Mortar B.S. 12 (1947)				
Mean strength (MN/m^2)	6.1	34.8	50.7	69.4
Mean strength $(lb/in.^2)$	890	5050	7350	10030
Standard deviation (MN/m^2)	3.2	5.0	3.9	4.4
Standard deviation $(lb/in.^2)$	460	720	570	640
Concrete with Water-Cement of 0.6 (1:1.5:3)				
Mean strength (MN/m^2)	15.4	30.7	39.7	57.4
Mean strength $(lb/in.^2)$	2230	4450	5760	8330
Standard deviation (MN/m^2)	4.6	3.9	4.8	4.6
Standard deviation $(lb/in.^2)$	670	560	700	660

[a] See reference 66.

Table 5.7 Compressive Strength of Concretes
(American Standards)[a]

Water-Cement Ratio		Compressive Strength Cured 28 days (lb/in.^{2})	
U.S. Gallon per Sack[b]	By Weight	Nonair Retained	Air Retained
4.00	0.35	6100	5000
5.00	0.44	5000	4000
5.16	0.46	4800	3900
6.00	0.53	4000	3200
6.44	0.57	3600	2900
7.00	0.62	3200	2600
7.74	0.69	2700	2200
8.00	0.71	2550	2050
9.00	0.80	2050	1650

[a] See reference 69.
[b] 1 U.S. sack of cement = 94 lb, 1 U.S. gallon of water = 8.33 lb.

The tension strength of concrete is about 0.1 of the compressive values, and tension strength in bending (rupture strength) is about 0.15 times compressive strength.

5.4   APPLICATIONS OF CONCRETE IN MINES

5.4.1   Shotcreting

"Shotcrete," also referred to as "gunite," is pneumatically applied mortar or concrete. It is defined as mortar or concrete that has been conveyed from the delivery equipment (generally called the "gun"), through a hose and pneumatically projected at high velocity onto a surface. A relatively dry mixture is generally used so that the material is capable of supporting itself without sagging or sloughing, even for vertical and overhead applications [70, p. 1-2].

The two basic shotcreting processes are (1) the widely used dry-mix process, where a mixture of cement and damp sand is conveyed through the delivery hose to a nozzle where the remainder of mixing

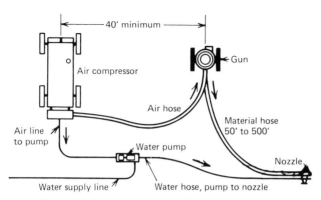

Figure 5.5 Typical dry shotcreting arrangement [70].

water is added; (2) the recently introduced wet-mix processes where all the ingredients (including water) are mixed before they enter the delivery hose. Either method will produce a quality of shotcrete suitable for normal requirements.

A typical arrangement for the dry process is shown in Fig. 5.5 [70, p. 17]. It consists of the gun, an air compressor, material hose, air and water hoses, nozzle, and sometimes a water pump.

First, the materials must be batched, usually in quantities of approximately 43 kg of cement to 180 kg of sand. The quantities are controlled both by volume and by weight. After batching, the mix is prepared by a drum mixer or in some cases by a screw mixer or conveyor. The mixed material in suspension is forced by compressed air through a hose to the nozzle. At the nozzle water is injected into the material in a number of fine streams. As the material passes through the 20–30-cm nozzle, it is mixed with the water. Mixing continues as the stream of material and water passes between the nozzle and the point of impingement. Upon impact, water mixing is complete.

A typical double-chamber gun is shown in Fig. 5.6 with the lower cone valve closed and the machine feeding the dry mix from the pressurized lower chamber recharged. The upper cone valve is then closed and the auxillary air valve opened. The pressure at the lower chamber is about $3.6–7 \text{ kg/cm}^2$.

The operation of a modern wet-mix shotcrete gun of the pure pneumatic feed type, is shown in Fig. 5.7 [70, p. 63]. The mixing chamber pictured is discharged under air pressure through a pneu-

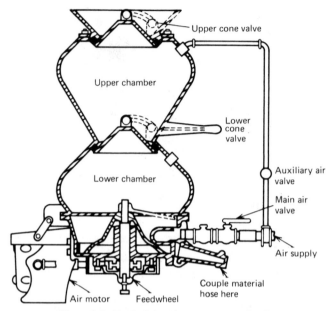

Figure 5.6 Typical double-chamber gun [70].

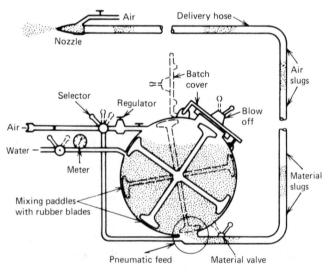

Figure 5.7 Schematic drawing of a wet-mix shotcrete gun of pure pneumatic type [70].

matic feed sump at the bottom of the mixer. The rate of discharge and velocity of flow through the delivery pipe are controlled jointly by the air pressure and the rotation speed of the mixing paddles. As each slug of material is discharged into the sump, it is forced by the bottom air pressure through a slightly restricted opening into the delivery hose at high velocity. The material is thus transported as alternate slugs of compressed air and material to the nozzle. At the nozzle more compressed air is added through a special air ring, which breaks up the slugs and gives added velocity to the material as it is gunned from the nozzle.

The granulometry of the shotcretes is shown in Fig. 5.8 for coarse and fine grouting [70, p. 38]. Line B and the shaded area around it

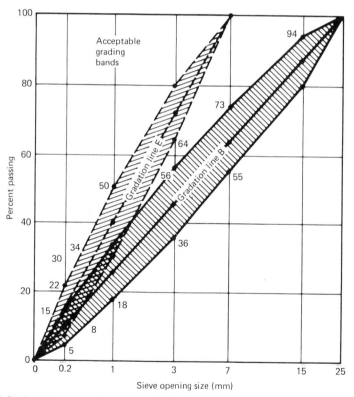

Figure 5.8 Granulometry of shotcretes [70]. Line B represents fine-grained material; line E, coarse-grained.

represents fine shotcreting, line E and related area for coarse grouting. In the dry-mix coarse-grained process the water-cement ratio is the easiest factor to control and ranges from 0.32 to 0.40, depending on the size, gradation, and quality of aggregates used. The cement content may range from 300 kg/m³ to 400 kg/m³ for shotcretes of 375–425 kg/cm² strength. In fine-grained shotcretes up to 500 kg/cm² strength is required.

One of the advantages of shotcrete is that it can be installed as soon as the mining openings are made, before the strata sags. After the explosive smoke clears, the shotcrete crew may come and gunite the roof, even before the muck is removed. After mucking the sides, the entire section of the gallery can be finished by shotcreting.

5.4.2   Monolithic Concreting

Monolithic concreting involves placing a 40–60-cm wall around the gallery roof and sides. Sometimes, where more pressure noticed, the entire periphery is concreted, as shown in Fig. 5.9 [2, p. 673]. The concrete is held in place by folds, usually made of iron sheets. It takes 2–4 weeks for good curing inside the folds.

The pit-bottom roadways and places of high pressure are first held by rigid arches, then the arches and the ground behind them are concreted, forming a reinforced concrete as shown in Fig. 5.10 [2, p. 678]. Such galleries can stand 25–30 years without any alterations.

5.4.3   Gallery Lining with Concrete Blocks

In places where movements of the strata are observed, monolithic concrete breaks or sloughs off, causing accidents. In such places,

Figure 5.9   Monolithic concreting of galleries [2].

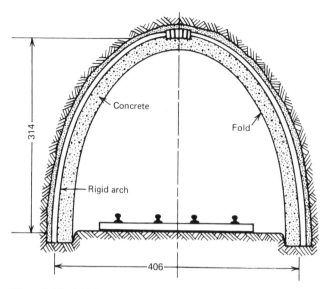

Figure 5.10 Rigid arches left in concrete for long-life galleries [2].

conical blocks are set in an arch form, and wooden blocks, 5-cm thick, are inserted between blocks to absorb the movement. The cross section and side elevation of such supports are shown in Fig. 5.11a and b. The roof is held temporarily by wooden supports such as r_1, r_4, r_5, r_8. Blocks shown as 5 are put in place by the help of a lift 3 and 4. Dimensions are given in Fig. 5.11c. The cross section of the gallery in the finished form is shown in Fig. 5.11d. The temporary wooden supports are left in place and extra concrete is added to the roof [71, 2, p. 676].

5.4.4 Concrete Shaft Lining

The support function in shaft sinking is an important matter and a most time-consuming operation. Here, only the high points are given briefly. The supporting system is explained schematically.

The shaft is sunk either by wooden support or channel profiles, temporarily. The shaft is sunk to a depth of 15–40 m, according to the rock conditions. Then, an "identation" is made, as shown in Fig. 5.12a [30, 2, p. 680], and monolithic concrete is poured, using iron

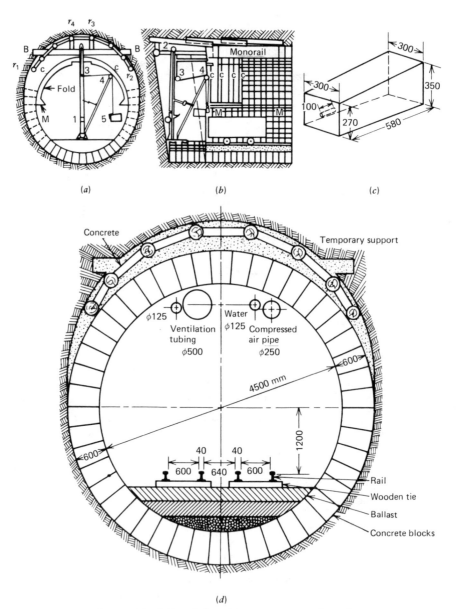

Figure 5.11   Gallery lining with concrete blocks [2, 71].

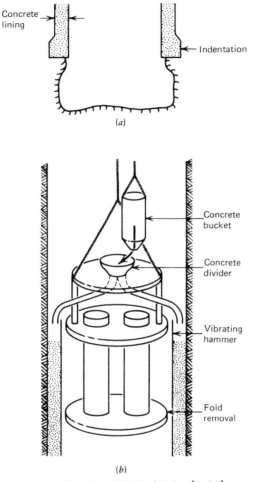

Figure 5.12 Monolithic shaft lining [2, 30].

folds, usually 3–4 m in depth. The concrete is prepared at the surface and is brought down by special buckets that open at the bottom, or by a pipe. "Moving platforms," made of two decks, are used for concreting and removal of the forms. The concrete is compacted by vibrators on every pouring, Fig. 5.12*b* [30, 2]. The design of the indentation and thickness of the shaft lining are dealt with in Section 5.5, on design.

5.4.5   Artificial Roofs

The use of an "artificial" roof in thick seams and lens type of metal-lic ore deposits is becoming popular. The ore is taken in slices in descending order [72]. The schematic view of such system to be adapted to a copper mine is illustrated in Fig. 5.13 [73, p. 20]. The panel is developed by two raises F_1, F_2, and each slice is mined by a central entry and "rooms" (1–20) leaving pillars between. Then the pillars are removed in second cycle of mining. A "roof" is made for the next slice by laying iron bars on the floors of the rooms, as shown in Fig. 5.14, and all installing a layer of concrete about 30 cm thick. Then the rooms are hydraulically stowed. The height of each slice is 2.5–3.0 m as shown in the panel section of Fig. 5.13. The design of such an "artificial roof" is given in Section 5.5.5.

5.5   DESIGN CONCRETE

5.5.1   Design in Concrete Preparation

The following two factors should be met in preparing concrete underground:

1.   It should have the compressive strength required.
2.   It should be fluid enough to be transported in a pipeline.

In pipeline transportation the slump is 7.5 ± 2.5 cm and the maxi-mum size of the aggregate is limited to 30 mm. Thus the cement, aggregate, and water should be calculated and the concrete formed should meet the characteristics required.

First, the average compressive strength should be evaluated. This should be [68, 69]

$$\sigma_{av} = \frac{\sigma_p}{1 - tV} \tag{5.10}$$

where σ_{av} = the average strength upon which the calculations are made
σ_p = project strength (specified strength)

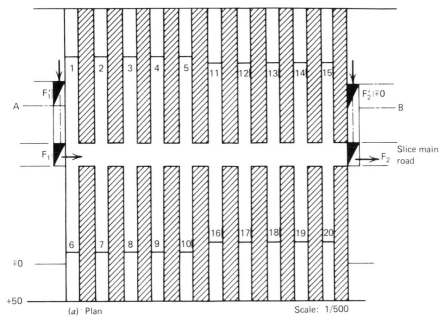

(a) Plan   Scale: 1/500

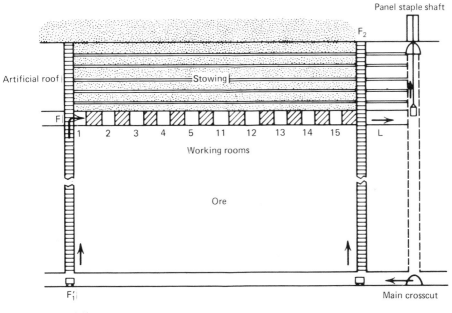

(b)

Figure 5.13 A typical panel working with descending slices using an artificial roof [73].

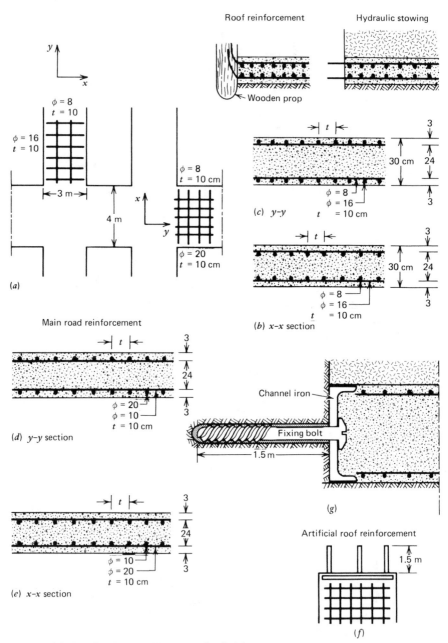

Figure 5.14 Details of the artificial roof [73]. (*a*) Panel plan; (*b*) *x-x* section; (*c*) *y-y* section; (*d*) *y-y'* section; (*e*) *x-x* section.

194

V = variation coefficient related to working conditions; in well controlled conditions, 0.15–0.10; in medium conditions, 0.15–0.20; and in poor conditions where quality cannot be controlled, 0.20–0.30

t = statistical coefficient; if 95% of the specimens should be accepted, $t = 1.645$, if 90% acceptable $t = 1.282$

Second, the α = water-cement ratio is calculated using several formulas (for example, the Bolomey formula). As the dosage is given at the beginning, the amount of water is calculated. If the characteristics of the aggregates are known, the amount of water is checked with the fluidity formula (5.2). In fluid concrete, more water is required.

Third, the total aggregate volume is calculated. As the cement, water and air are known, they are subtracted from 1m^3, and the rest is the aggregate (sand-gravel). An assumption is made for gravel-sand ratio, then their volumes are found respectively.

As explained, the preparation of the concrete relies on many assumptions. Therefore samples should be prepared and unit weight, compactness, and fluidity determined. Then samples should be tested upon completion of their curing times. If the strengths determined by testing do not meet the project requirements, some corrections should be made. These corrections may be to the granulometry of the aggregates, amount of cement, and amount of water. The most difficult correction involves the granulometry. In practice, many failures are due to this factor.

As a numerical example let us calculate a monolithic shaft lining, for which the following data are given:

Compressive strength of concrete after 28 days	160 kg/cm^2
Conditions of concrete	Fluid
Aggregate and maximum size	Gravel, 25 mm
The fineness modulus	$K = 3.1$

Let us take good working conditions ($V = 0.15$) with 95% of specimens acceptable ($t = 1.645$). Then the average strength according to Eq. (5.10) is

$$\sigma_{av} = \frac{160}{1 - 1.645 \times 0.15} = 212 \longrightarrow 225 \text{ kg/cm}^2$$

We first calculate the α factor by the Bolomey Eq. (5.4)

$$\sigma_{av} = 180 \left(\frac{1}{\alpha} - 0.5 \right) \text{ kg/cm}^2$$

$$225 = 180 \left(\frac{1}{\alpha} - 0.5 \right)$$

$$\alpha = 0.57$$

If the dosage of $M_c = 350$ kg/m^3 is taken, the amount of water is

$$M_w = 350 \times 0.57 = 200 \text{ kg/m}^3$$

The fluidity condition should be checked. According to Eq. (5.2), for the given fineness modulus 3.1,

$$M_w = 58(7 - 3.1) = 226 \text{ kg/m}^3$$

There is not much difference between the two quantities of water determined, so the first one is taken for calculations. If γ_c and γ_a are the densities of cement and gravel,

$$\frac{M_c}{\gamma_c} + \frac{M_a}{\gamma_a} + M_w + V_{air} = 1 \text{ m}^3$$

The amount of air is taken as 1% and the amount of aggregates is given by

$$\frac{350}{3.11} + \frac{M_a}{2.65} + 200 + 10 = 1000 \text{ liters}$$

$$M_a = 1797 \longrightarrow 1800 \text{ kg/m}^3$$

To summarize all the constitutents for 1 m^3 of concrete are listed as follows:

Cement	$M_c = 350$ kg/m^3
Water	$M_w = 200$ kg/m^3
Total aggregate	$M_a = 1800$ kg/m^3

These figures are not final. Upon testing, corrections should be made if the results do not give compressive strength of 225 kg/cm^2.

5.5.2   Design for Shotcreting

According to Rabcewicz [74] the following formula* is given for shotcretes of normal conditions. The results agree quite well with those at other investigators.

$$t = 0.434 \frac{Pr}{\tau}$$ (5.11)

where t = thickness of shotcrete, in meters
P = stress on the shotcrete in tonnes per square meter
r = radius of the gallery, in meters
τ = allowable shear stress of the shotcrete material

Typical applications for main roadways 4.25 m wide and pit-bottom galleries 6.0 m in width in a copper mine are given in Fig. 5.15 [73, p. 7].

The pressure on the galleries is evaluated as 15 t/m², which is quite acceptable in coal mines as well. The shear strength of shotcrete is $0.2 \, \sigma_b$ (compressive strength), and this can be taken as 225 kg/cm² or 2250 t/m². If we assume a safety factor of 3, the allowable shear strength is as follows:

$$\tau_{sf} = \frac{0.2 \, \sigma_b}{F} = \frac{0.2 \times 2250}{3} = 150 \text{ t/m}^2$$

For galleries of r_1 = 2.125 and r_2 = 3.0 m

$$t_1 = 0.434 \times \frac{15 \times 2.125}{150} \cong 0.10 \text{ m}$$

$$t_2 = 0.434 \times \frac{15 \times 3.0}{150} = 0.13 \text{ m} \longrightarrow 0.15 \text{ m}$$

In Sweden the standard for shotcrete thickness is 10–15 cm for fissured ground. According to Wickham and Tieman [75], 12.5 cm at roof and 6.5 cm on the sides are quite sufficient. Deere [76] has shown that 3 cm of shotcrete in a gallery 4 m wide can support pyramidal blocks 1 × 1 × 1 m (0.9 t) at a safety factor of 13, as shown in Fig. 5.16 [76, 77, p. 3–38].

*Safety factor F in design is taken to be 1.

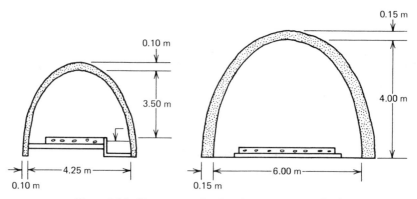

Figure 5.15 Shotcrete applications in a copper mine [73].

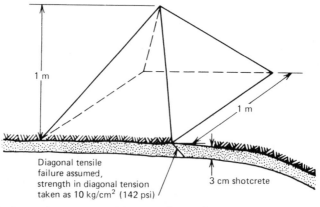

Figure 5.16 Supporting capacity of shotcrete [76, 77]. Weight of pyramid = 900 kg. Shotcrete resistance = 4 X 1 m X 3 cm X 10 kg/cm^2 = 12000 kg. Factor of safety against fallout = 12000/900 = 13.

5.5.3   Design of Shaft Lining

As is quite difficult to calculate the thickness of lining analytically, many approximate formulas are used for practical work. The following equations apply to Fig. 5.17 [2, 689].

According to Protodjakonov [78]

$$t = \frac{Pr}{(\sigma_b/F) - P} + \frac{150}{(\sigma_b/F)} \tag{5.12}$$

$$t = 0.007\sqrt{2rH} + 14 \tag{5.13}$$

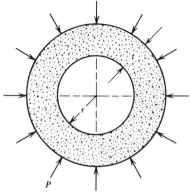

Figure 5.17 Shaft lining and side pressures [2].

According to Brinkhaus [79]

$$t = \frac{2r}{10} + 12 \tag{5.14}$$

According to Heber [80]

$$t = \left(\sqrt{\frac{\sigma_b/F}{\sigma_b/F - 2P}} - 1 \right) r \qquad \text{(medium, } < 400 \text{ m deep) (5.15)}$$

$$t = \left(\sqrt{\frac{\sigma_b/F}{\sigma_b/F - \sqrt{3}P}} - 1 \right) r \qquad \text{(deep} > 400 \text{ m)} \tag{5.16}$$

where t = thickness of lining, in centimeters
P = side pressure on lining, in kilograms per square centimeter
H = depth of shaft from surface, in centimeters
r = radius of shaft, in centimeters
σ_b = 28-day strength of concrete, in kilograms per square centimeter
F = safety factor, usually 2

As a numerical example to show the usage of these equations the following data are given:

Depth of shaft	$H = 300$ m
Formation	Sandstone
Bearing capacity	$\sigma_{\text{safe}} = 15$ kg/cm^2
Poisson's number	$m = 5$
Density	$\gamma = 2.5$ t/m^3

Concrete used
   28-day compressive strength $\sigma_b = 225$ kg/cm^2
   Density $\gamma_b = 2.4$ t/m^3
Radius of shaft $r = 2.5$ m
   Safety factor $F = 2$

Let us first calculate the horizontal stress to the shaft lining.

$$P = \frac{0.1\,\gamma H}{m - 1} = \frac{0.1 \times 2.5 \times 300}{5 - 1}$$

$$= 18.75 \text{ kg/cm}^2$$

Assuming σ_b 225 kg/cm^2 as the σ_t strength of lining,

$$t_1 = \frac{18.75 \times 250}{(225/2) - 18.75} + \frac{150}{(225/2)}$$

$$= 51.33 \text{ cm}$$

$$\cong 52 \text{ cm}$$

$$t_2 = \left(\sqrt{\frac{225/2}{225/2 - 2 \times 18.75}} - 1 \right) 250 = 56 \text{ cm}$$

$$t_3 = 0.007\sqrt{2 \times 250 \times 30000} + 14 = 41 \text{ cm}$$

$$t_4 = \frac{500}{10} + 12 = 62 \text{ cm}$$

It can be seen that the results of 52, 56, 41, and 62 cm are in close accordance with each other.

5.5.4   Design of Shaft Indentation

When installing the lining, a small "indentation" is made all around the shaft to transfer the dead weight of the lining to the main rock. This is repeated for every pouring of concrete, usually at 20–40 m intervals. The indentation is shown in Fig. 5.18 [2, p. 695], and formulas are given as follows:

$$a \geqslant \sqrt{(r + t)^2 + \frac{(2r + t)\,th\gamma_b\,\cos^2\alpha}{\sigma_{sf}}} - (t + r) \quad (5.17)$$

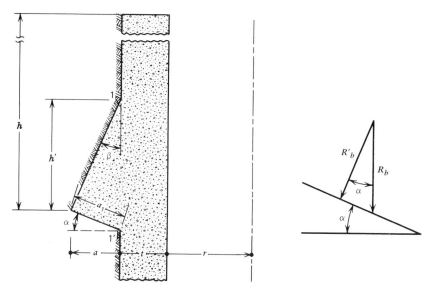

Figure 5.18 Design of shaft identation [2].

$$h' \geqq \frac{(2r + t) \, th\gamma_b}{2(r + t) \, \tau_{sf}} \tag{5.18}$$

$$\tan \beta = \frac{a}{h' - a \tan \alpha} \tag{5.19}$$

where a = thickness of indentation, in centimeters

r = inner radius of shaft, in centimeters

t = thickness of shaft lining, in centimeters

h = height of the lining to be poured, in centimeters

h' = height of indentation, in centimeters

γ_b = density of the concrete, in kilograms per square centimeter

σ_{sf} = bearing capacity of rock, in kilograms per cubic centimeter

τ_{sf} = safe shear strength of concrete, in kilograms per square centimeter

α = horizontal angle of indentation, in degrees

β = vertical angle of indentation, in degrees

Taking the thickness of the preceding example as 50 cm, the height of concrete as 40 m, and the angle of indentation as $\alpha = 0°$, the indentation depth a is:

$$a \geqslant \sqrt{(250+50)^2 + \frac{(2 \times 250 + 50)(50)(4000)(0.0024)}{15}}$$

$$- (250 + 50)$$

$$= 28 \text{ cm}$$

The safe shear strength of concrete may be calculated as follows:

$$\tau_{sf} = 0.5\sqrt{\sigma_b} = 0.5\sqrt{225} = 7.5 \text{ kg/cm}^2 \tag{5.20}$$

$$h' = \frac{(2 \times 250 + 50)(50)(4000)(0.0024)}{(2)(250+50)(7.5)}$$

$$= 58 \text{ cm}$$

These values, although safe, are small for practical use. The depth is assumed to be 1.5 times the thickness of the shaft lining, and the vertical angle is taken to be $\beta = 15\text{–}20°$. Then,

$$a = 1.5(50) = 75 \text{ cm}$$

$$h' = \frac{a}{\tan \beta} = \frac{75}{\tan 20°} \cong 200 \text{ cm}$$

5.5.5 Design of Artificial Roofs

The design of iron bars for reinforced concrete is part of the detailed work of civil engineering. Reference 81 describes such a project to be used in a lignite mine. In the example given in Figs. 5.13 and 5.14 the average pressure is calculated using Terzaghi's formula, Eq. (1.34). The pressure is found to be 10 t/m², which is quite acceptable for many design concepts.

In the example of Figs. 5.13 and 5.14 the rooms are 3.0 m wide. The thickness of the reinforced concrete is taken as 30 cm. The yield strength of 2200 kg/cm² at a safety factor of 1.5 is assumed for iron bars. The concrete is prepared according to a 28-day cured strength of 160 kg/cm² at safety factor 1.15. The maximum moments are calculated and the spacings of bars are found and shown as in Fig. 5.14. On rooms (sections b, c) 8- and 16-mm diameter bars are spaced at 10 cm; on the main roadway (sections d and e),

Table 5.8 Materials Used in Artificial Roofs of Reinforced Concrete[a]

Material (kg)	Per 1 m^3 of Concrete	Per Ton of Ore
Iron bars	176.3	7.05
Cement	320	12.8
Sand (0–7 mm)	702	28.1
Broken rock (7–25 mm)	1098	43.9
Water	197	7.7

[a]See reference 73.

bars 20 and 10 mm in diameter spaced at 10-cm intervals are calculated. The roof is anchored at the boundary to the host rock by roof bolts as shown in Fig. 5.14f and g.

The concrete material, to be transported from a mixing plant on the surface by a pipeline, is constituted of aggregate smaller than 25 mm, dosage of 300 kg/m^3, compaction of 0.85 and water-cement ratio of $\alpha = 0.6$. The amounts of material per cubic meter of concrete and per ton of ore are as given in Table 5.8 [73, p. 36–37].

APPENDIX 5.1   RELEVANT BRITISH STANDARDS

A. Cement

B.S.   12: 1958   Portland cement (ordinary and rapid-hardening) (metric version, 1971)
       146: 1958   Portland-blast furnace cement (metric version, 1973)
      1370: 1958   Low heat Portland cement (metric version, 1974)
      4246: 1968   Low heat Portland-blast furnace cement (metric version, 1974)
      4248: 1974   Supersulphated cement
       915: 1947   High alumina cement (metric version, 1972)
      1014: 1961   Pigments for cement, magnesium oxychloride and concrete

B. Aggregate

B.S. 882, 1201: 1965	Aggregates from natural sources for concrete (including granolithic)
812: 1967	Methods for the sampling and testing of mineral aggregates, sands, and fillers
877: 1967	Foamed or expanded blast furnace slag lightweight aggregate for concrete (metric version, 1973)
1047: 1952	Air-colled blast furnace slag coarse aggregate for concrete (metric version, 1974)
1165: 1966	Clinker aggregate for concrete
410: 1969	Test sieves
3797: 1964	Lightweight aggregates for concrete
3681: 1963	Methods for sampling and testing of lightweight aggregates for concrete (metric version, 1973)

C. Concrete

B.S. 1881: Part 1: 1970	Methods of sampling fresh concrete
1881: Part 2: 1970	Methods of testing fresh concrete
1881: Part 3: 1970	Methods of making and curing test specimens
1881: Part 4: 1970	Methods of testing concrete for strength
1881: Part 5: 1970	Methods of testing hardened concrete for other than strength
1881: Part 6: 1971	Analysis of hardened concrete
4408: Part 1: 1969	Electromagnetic cover measuring devices
4408: Part 2: 1969	Strain gauges for concrete investigations
4408: Part 3: 1970	Gamma radiography of concrete
4408: Part 4: 1971	Surface hardness methods
4408: Part 5: 1974	Measurement of the velocity of ultrasonic pulses in concrete
1926: 1962	Ready-mixed concrete
1305: 1974	Batch type concrete mixers
3963: 1965	Method for testing the performance of batch type concrete mixers
368: 1971	Precast concrete flags
2028, 1364: 1968	Precast concrete blocks
3148: 1959	Tests for water for making concrete

APPENDIX 5.2   SELECTED LIST OF RELEVANT ASTM STANDARDS*

A. Cement

C 150–74	Spec. for Portland cement
C 595–74	Spec. for blended hydraulic cements
C 115–74	Test for fineness of Portland cement by the turbidimeter
C 186–73	Test for heat of hydration of hydraulic cement
C 151–74a	Test for autoclave expansion of Portland cement

B. Admixtures

C 618–73	Spec. for fly ash and raw or calcined natural pozzolans for use in Portland cement concrete
C 494–71	Spec. for chemical admixtures for concrete
C 441–69	Test for effectiveness of mineral admixtures in preventing excessive expansion of concrete due to the alkali-aggregate reaction
C 260–73	Spec. for air-entraining admixtures for concrete

C. Aggregate

C 294–69	Descriptive nomenclature of constituents of natural mineral aggregates
C 33–74	Spec. for concrete aggregates
C 330–69	Spec. for lightweight aggregates for structural concrete
C 331–69	Spec. for lightweight aggregates for concrete masonry units
C 332–66	(1971) Spec. for lightweight aggregates for insulating concrete
C 117–69	Test for materials finer than No. 200(75-μm) sieve in mineral aggregates by washing
C 70–73	Test for surface moisture in fine aggregate
C 40–73	Test for organic impurities in sands for concrete
C 123–69	Test for lightweight pieces in aggregate
C 88–73	Test for soundness of aggreates by use of sodium sulfate or magnesium sulfate

*T denotes Tentative Standard. The two digits after the dash denote the year of publication.

C 131–69   Test for resistance to abrasion of small size coarse aggregate by use of the Los Angeles machine

C 289–71   Test for potential reactivity of aggregates (chemical method)

C 227–71   Test for potential alkali reactivity of cement–Aggregate combinations (mortar-bar method)

C 586–69   Test for potential alkali reactivity of carbonate rocks for concrete aggregates (rock cylinder method)

C 638–73   Descriptive nomenclature of constituents of aggregates for radiation-shielding concrete

C 637–73   Spec. for aggregates for radiation-shielding concrete

E 11–70   Spec. for wire-cloth sieves for testing purposes

D. Concrete

C 124–71   Test for flow of Portland cement concrete by use of the flow table (discontinued 1974)

C 143–71   Test for slump of Portland cement concrete

C 360–63   (1968) Test for ball penetration in fresh Portland cement concrete

C 403–70   Test for time of setting of concrete mixtures by penetration resistance

C 232–71   Test for bleeding of concrete

C 138–74   Test for unit weight, yield, and air content (gravimetric) of concrete

C 173–73a   Test of air content of freshly mixed concrete by the volumetric method

C 231–73   Test for air content of freshly mixed concrete by the pressure method

C 470–73T   Spec. for molds for forming concrete test cylinders vertically concrete test cylinders

C 192–69   Making and curing concrete test specimens in the laboratory

C 39–72   Test for compressive strength of cylindrical concrete specimens

C 617–73   Capping cylindrical concrete specimens

C 78–64   (1972) Test for flexural strength of concrete (using simple beam with third-point loading)

C 496–71   Test for splitting tensile strength of cylindrical concrete specimens

C 42–68	(1974) Obtaining and testing drilled cores and sawed beams of concrete
C 215–60	(1970) Test for fundamental transverse, longitudinal, and torsional frequencies of concrete specimens
C 418–68	(1974) Test for abrasion resistance of concrete
C 85–66	(1973) Test for cement content of hardened Portland cement
C 457–71	Rec. practice for microscopical determination of air-void content and parameters of the air-void system in hardened concrete
C 666–73	Test for resistance of concrete to rapid freezing and thawing
C 94–74	Spec. for ready-mixed concrete
C 156–74	Test for water retention by concrete curing materials

CHAPTER 6
Stowing

6.1 IMPORTANCE OF STOWING

The term "stowing" includes all the steps taken to "fill" the openings made by extraction of the seams of mineral deposits. It is a part of the support system. If stowing follows right after the excavations, it diminishes the movement of strata and helps the roof and surface control tremendously.

The "room and pillar" system of mining, with proper size of pillars, is quite effective in surface control. However, in deeper mining the size of pillars increases, and the percentage of extraction of minerals diminishes. Moreover, the pillars may cause difficulties by deteriorating, cracking, catching fires, and so on.

"Caving systems," especially with powered supports are very economical and fast, providing high production. But, the material produced entails much expense to compensate for surface damages. Besides, it is possible for seas, lakes, rivers, canals, and other surface features to be disturbed, causing flooding of the mine and the extra expense of pumping out the water. There are seams with strong roof layers where the caving is difficult. In such cases caving is achieved by "weighting," causing damage to the supporting elements.

The advantages to stowing systems is that they minimize the surface disturbance. Since an area is "filled up" as soon as an opening is made, the main roof does not sag or cause excessive weighting. In this respect strata control is the most easily achieved and the most effective.

6.1.1 Amount of Stowing Materials

The following formula is utilized to calculate the weight of stowing material:

$$\frac{P'}{\gamma'} = \frac{P}{\gamma} K \tag{6.1}$$

where P = weight of mineral extracted, in tonnes
$\quad P'$ = weight of stowing materials to be used, in tonnes
$\quad \gamma$ = density of ore or coal, in tonnes per cubic meter
$\quad \gamma'$ = density of stowing materials, in tonnes per cubic meter
$\quad K$ = factor of stowing (0.3–0.95) according to stowing systems

If we apply this equation to coal and metallic deposits where γ is 1.3 and 3.0 t/m^3, respectively, γ' is about the same for both cases, which can be taken 1.6 t/m^3. The K factor for pneumatic stowing in coal mines is about 0.8, in metal mines 0.7. So the weight of stowing material is

$$P' = \frac{\gamma'}{\gamma} KP$$

$$= \frac{1.6}{1.3} \times 0.8 \times P \cong P' \qquad \text{(coal)}$$

$$P' = \frac{1.6}{3.0} \times 0.7 \times P \cong 0.4 P \qquad \text{(metal)}$$

It can be seen that the amount of stowing material for a coal mine equals the daily production. It is a difficult task to prepare such a large amount and haul it to the place of production using haulage facilities on "upgrade" incline. However, in metal mines, the work is easier to manage owing to small daily production and a small ratio (0.4).

6.1.2   Sources of Stowing Materials

As the amount of stowing material is voluminous, all available sources are used to obtain this large quantity. These sources are summarized in the following sections.

Goaf Stone.   The fallen part of the roof is a ready-made material that can be utilized with minimum transportation distance. This is used as "packing" material for chocks left to support gateways. The stone obtained from goaf is packed inside the chocks and in the space between chocks.

Formerly "strip packing" was used to support the roof as the face advanced, utilizing the goaf stone. However, this source is very scarce and not reliable. In metal mines small raises of 40° are driven to obtain stowing material in the stope.

Gateway Brushings.   The brushing of gateways in thin seams produces stone that can be easily utilized in packing gateway sides. A small crusher and a pneumatic machine can do the work very effectively.

Development Work of the Mine.   The stone resulting from development work, like staple shafts, crosscuts, and other entries, is a good source for as much as 25% of stowing material. This stone must be crushed to a size of - 80 mm to be used as stowing material. Crushers can be installed on every level to eliminate hoisting, or a central crushing plant can be installed at the surface to serve the entire mine.

Washer-Concentrator Rejects.   Rejects form the most important source of stowing material in respect to quantity and quality. All the rejects can be utilized in mixing with other materials. In hydraulic stowing, materials smaller than 0.1 mm can cause trouble, so these should be separated from the other rejects. Such a utilization eliminates the cost of handling and heaping such materials at the surface and helps to meet environmental regulations.

Old Refuse Piles.   Refuse from old mines can be utilized in the same way as washer-concentrator rejects, provided screening and crushing facilities are used to separate suitable sizes for stowing.

Quarries.   If the preceding sources are not sufficient, a quarry may be opened and the stone obtained utilized after crushing. A conglomerate formation is best suited for stowing. River beds are also good sources of stowing material, as the material available is free of small particles.

6.1.3   Advantages of Stowing

1. Stowing is necessary in mine seams steeper than 45°. Seams can be easily mined by changing the incline to 42° (diagonal faces) by advancing with "gravity" stowing.

2. In stowing systems the pressure arch can be small as shown in Fig. 6.1. So, the abutment pressures on the gateways are much less, and the supporting problems of the mine are much easier.

3. Stowing adds to safety as the roof does not break; roof falls and accidents due to falls are minimized.

4. Subsidence has been minimized. Especially, sea-river-canal-lake-site extractions are made possible by stowing systems.

5. Seams with strong roofs can be mined safely be stowing systems, eliminating excessive weighting.

6. Refuse piles and their dangers of sliding and pollution are

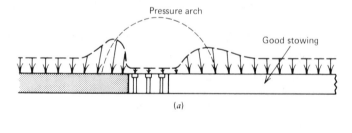

(a)

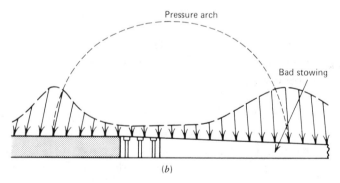

(b)

Figure 6.1 Pressure arch in stowed longwalls [2]. (a) Small pressure arch; less stress, good roof control. (b) Large stresses, bad roof control.

eliminated. Disfiguring the landscape is prevented, helping to meet the environmental regulations.

6.1.4   Disadvantages of Stowing

1. The greatest disadvantage to stowing is the increase in cost it causes. This may be twofold at the face but diminishes, taking into consideration the other expenses, especially the refuse handling. This is discussed in Section 6.4.
2. Large capital investment is required. A large plant and pipelines must be installed underground for hydraulic stowing. Although the expense is less for pneumatic stowing, large compressed-air capacity is required.

6.2   APPLICATION OF STOWING SYSTEMS

Stowing applications differ according to the source of energy required to install the material and are classified according to those sources.

6.2.1   Hand Stowing

In times when mechanization was not advanced and labor costs were small hand stowing was used.

"Strip packing" is schemetized in Fig. 6.2a, b. Workers build dry walls and, as the packing progresses, throw small rocks by shovel behind the dry walls. The material is obtained from the goaf by partial caving of the immediate roof. If the materials are not adequate, blasted roof materials are added. The effect of the quality of the stowing is shown in Fig. 6.2e. In poor filling the great amount of pressure on the arches of the gateways causes deformations. Good filling diminishes such deformations, as shown in the figure.

In metal mines, entries made from the main entry of the panel are stowed on a retreating system as shown in Fig. 6.2c, d. The stowing material is brought from raises of the level above or obtained by small, 40°, raises driven in the footwall rock.

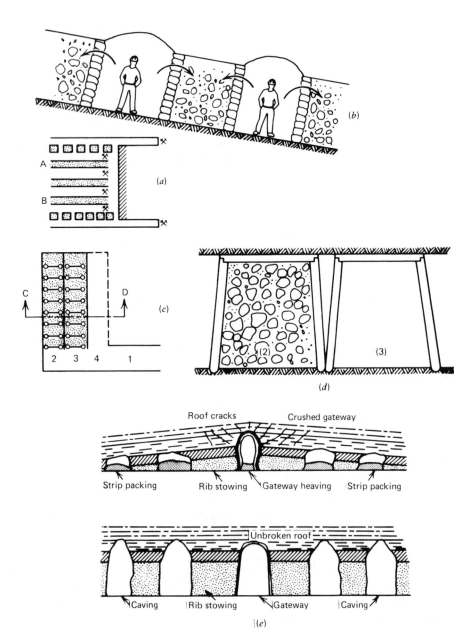

Figure 6.2 Hand stowing systems [2].

6.2.2 Gravity Stowing

The force of gravity can be used to place stowing material. This method is used in seams steeper than 42°, either diagonally or down the maximum inclination, as shown in Fig. 6.3a, b.

The stowing material is washery refuse mixed with broken mine rock, and it is disposed along an inclination of $\varphi = 42°$ (internal friction angle of rocks). The cross sections of the faces are shown in Fig. 6.3c, d. If the coal is stable enough, it is easier to work diagonal faces without "wire screens." Wedges (4) can hold the face coal (Fig. 6.3c). If the coal is unstable, K-type of support (5) is needed, and the stowing material is held in place by a "stowing screen (3)" behind the supports (Fig. 6.3d).

Another system is "stepped face" or *gradins reversés* (turned over stairway) as shown in Fig. 6.3e. Special wooden supports are used and left in the stowing material, one placed on top of the other, resembling "square setting." The stowing follows the stepped face, at a 42° inclination and a distance of 2–3 m.

6.2.3 Mechanical Stowing

In the mechanical stowing system, materials are delivered by conveyor and are thrown to the back of the face by a "jet conveyor," as shown in Fig. 6.4. The materials are brought by a conveyor. A diagonal scraper transfers the material to a small conveyor below, working at a speed of 10 m/s, thus throwing the materials to back of the face. The jet conveyor is pulled up slowly as the stowing progresses. A wire screen keeps the stowing away from the face line.

The system is adapted to thick and flat seams, as two conveyors require a height of more than 1.5 m. Therefore, where space is limited, it is replaced by pneumatic stowing which requires much less space.

6.2.4 Pneumatic Stowing

In pneumatic stowing the stowing materials are conveyed in pipes and thrown to the back of the face by compressed air. This is the most popular system of stowing because it requires smaller installa-

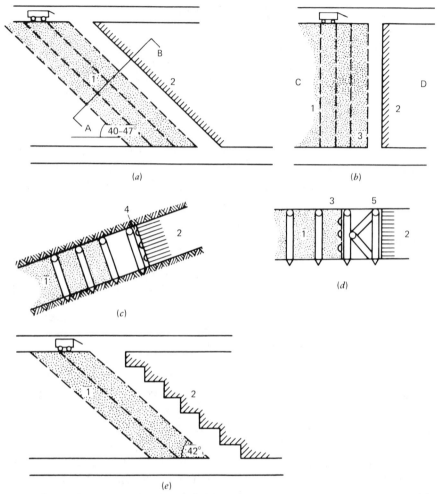

Figure 6.3 Gravity stowing systems [2]. (*a*) Diagonal seam; (*b*) maximum inclination; (*c*) A-B section; (*d*) C-D section; (*e*) stepped face.

tions. However, the mine should have an ample supply of compressed air, as the amount of air spent by one stowing machine is almost equivalent to a medium-sized compressor at the surface.

The applications of pneumatic systems in coal and metal mines are illustrated in Figs. 6.5 and 6.6, and stowing machines are shown in Fig. 6.7. Detailed descriptions of stowing can be found in references 82 and 83.

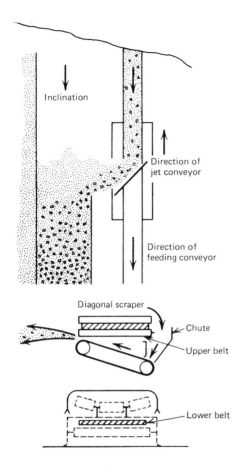

Inclination

Direction of
jet conveyor

Direction of
feeding conveyor

Diagonal scraper

Chute

Upper belt

Lower belt

Figure 6.4 Mechanical stowing system
[2, 30].

The schematic plan view shows stowing materials brought by cars and tippled at 10 (Fig. 6.5a); a chain conveyor takes them to the stowing machine (1). They are blown into the basalt-lined pipeline (2). The 90° elbow (3) changes direction down to the face. Manganese steel pipes of shorter lengths (4), easily dismantled, deliver the stowing materials to the back of the face at a high speed. The cross section of the face (Fig. 6.5b) shows the pipe (4), with stowing already in place (9) and screen or wire paper (5) nailed to the wooden posts.

Stowing in a metal mine is shown in Fig. 6.6 [83; 2, p. 719]. The rooms or entries are filled by stowing materials conveyed in a pipeline from the bottom of a stowing raise. High entries may require double-deck stowing as shown in the figure.

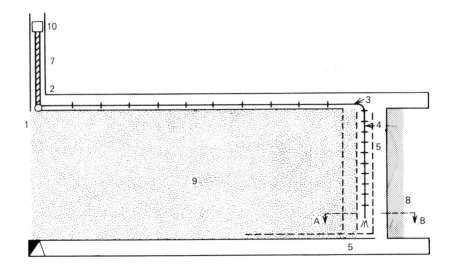

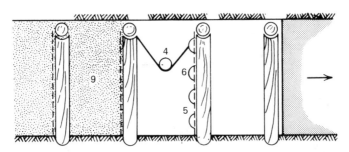

Figure 6.5 Pneumatic stowing in coal mines [2]. (a) Plan; (b) A-B section.

The side and elevations of stowing machines used in coal mines at high capacity (70–150 m³/h) are shown in Fig. 6.7a [44]. The smaller-capacity but longer-distance machine is shown in Fig. 6.7b. The upper machine has drum-type feeder while the lower is equipped with screw-type feeders. Both deliver material to the pipes by compressed air at a pressure of 5–7 kg/cm².

The pipes on tailing roadways are lined inside with basalt to minimize the wear and last for 500,000 tons of material. The manganese steel pipes at the face are worn by 300,000 tons and the 90° elbow by only 6000 tons. There should be direct telephone connections between the stowing machine and the pipe operator. In case of stop-

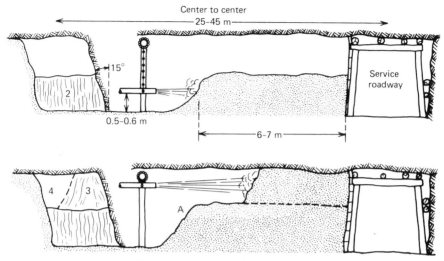

Figure 6.6 Pneumatic stowing in metal mines [2, 83].

pages, the materials are cut first and the air is kept blowing to eliminate settling in the pipes.

6.2.5   Hydraulic Stowing

The hydraulic stowing method is the most advanced filling system: stowing materials are mixed with water and delivered to the mine in a pipeline. It requires a preparation plant at the surface, pipelines, canals, sumps, and a pump room to raise the excess of water to the surface for reuse. The material should be small enough (- 80 mm) to be transported in a pipeline and large enough (>0.1 mm) not to stay in suspension in water. Otherwise, such material sinks in the canals and sumps and causes excessive cleaning expenses. The best material is river sand from which the slime has already flushed away by the river. The refuse of washeries, classified and separated from slime, is very satisfactory. The slag of smelters, suddenly cooled in water, is an excellent material, as well.

The use of an hydraulic stowing system to a steep coal seam is shown in Fig. 6.8 [2, p. 724]. The material is carried by the pipeline (3) and divided at the face every 15 m or so (4). A "dam" is made of jute curtain fixed on supports by wedges (5). The material

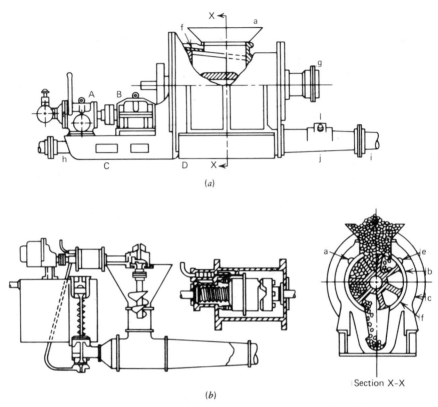

Figure 6.7 Drum (*a*) and screw (*b*) stowing machines [2, 44].

is deposited behind the dam, and water percolates through the pores of the jute curtain. The face operates horizontally, supported by wooden caps and posts in K-form (9, 10). The coal is kept in place by extra wedges (8, 6).

The plan and longitudinal section of a panel in a metal mine is shown in Fig. 6.9 [84; 2, p. 735]. The dams (4) are erected at the narrower sections of the stope. The transport of ore is done by scrapers to a central chute (5).

A typical surface plant installation is depicted in Fig. 6.10 [30; 2, p. 729]. The material excavated is dumped into a bunker and fed to a inclined screen. The finer material passes through the screen and coarser material is crushed by rolls. Oversided material is separated again by vibrator screens as shown in the figure.

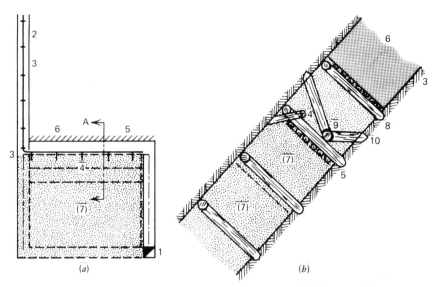

Figure 6.8 Schematic view of hydraulic stowing in a coal mine [2]. (a) Plan; (b) A-B section.

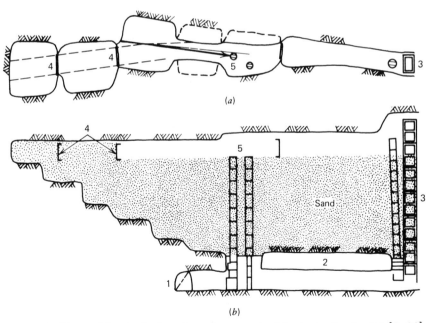

Figure 6.9 Plan and longitudinal section of stope, mined by hydraulic stowing [2, 84]. (a) Plan; (b) longitudinal section.

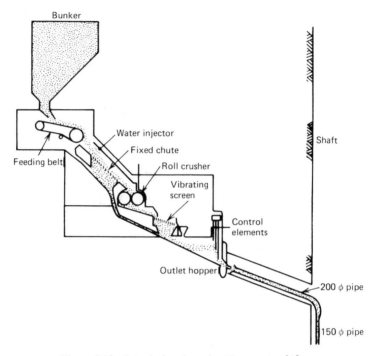

Figure 6.10 A typical surface plant for stowing [2].

All the fine material is mixed with an extra amount of water and sent to the mine. Care should be taken not to let any air bubbles enter the pipeline, as this air may cause trouble in the elbows.

6.2.6   Consolidated Stowing

Although the hydraulic stowing material is compacted after the water percolates through the pores of jute curtains, it is not solid. This material can be consolidated by adding special materials.

Sulphides are the easiest materials to be added to hydraulic stowing materials. Such sulphides oxidize in the stope, raising the temperature to 60–70°C and "cementing" the materials. In such a consolidated mass new development work such as raises and crosscuts can be driven. Pyrrhotite (FeS) is added (3%) in this way to the stowing material or suddenly cooled copper slags. The analyses of such materials are given in Table 6.1 [85; 2, p. 725].

Table 6.1 Analyses of Consolidated Stowing Materials[a]

Hydraulic Stowing Materials	Percent	Analysis of Pyrrhotite Minerals	Percent	Analysis of Slag Elements	Percent
Slag granulated in	73	Pyrrhotite	56	Fe	36.5
water		Pyrite	6	SiO_2	38.0
Crushed slag–100 mm	25	Magnetite	10	Al_2O_3	6.5
Pyrrhotite concentrate	3	Nonsoluble	28	S	1.5
	$\overline{100}$		$\overline{100}$	CaO	1.5
				MgO	1.0

Compressive strength of consolidated stowing: 20–70 kg/cm^2

[a]See references 2 and 85.

Low-dosage concrete is another consolidated stowing material (up to 20%). The amount of cement and the compressive strengths of such material are shown in Fig. 6.11 [86; 2, p. 737]. It can be seen that consolidated stowing made of sand and cement 5:1 can reach the compressive strength of 56 kg/cm^2 at about 18% cement content.

An application of such cemented consolidated stowing is illustrated in Fig. 6.12 [87; 2, p. 736]. The low-cement mixture is dumped by trucks, and water is added at the rate of 6–8 l/s. Such a mixture is delivered to the stope, where excess water is filtered and the material left consolidates in the stope.

6.3 DESIGN OF HYDRAULIC STOWING

Other stowing systems are quite simple and do not require much in the way of design. However, the hydraulic stowing system should be designed according to characteristics of the mine working, stowing materials available, and so on. The flow sheet of such a design is shown in Figure 6.13 [2, p. 747].

The design is better followed by a numerical example. Let us assume the following data:

Physical Data

Stowing material and density	Sandstone, $\gamma_k = 2.5$ t/m^3
Average size	$d = 2$ mm
Concentration of mixture	$K = 0.30$

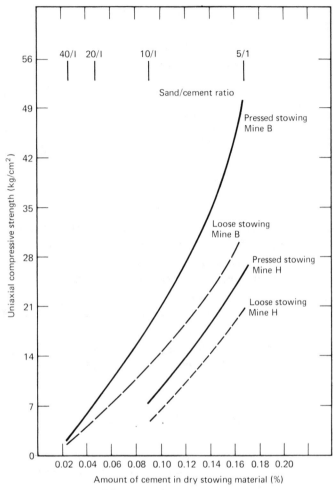

Figure 6.11 Cement-consolidated stowing material strengths [2, 86].

Technical Data

Dimension of the face   100 m × 1.2 m × 2 m
Duration of stowing   3 h/shift
Stowing shift   1 shift/day
Type of pipe   Steel

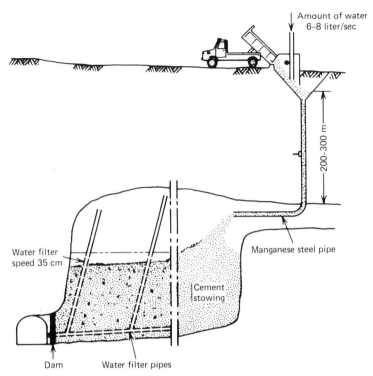

Figure 6.12 Consolidated-cement stowing system [2, 87].

Points	Geometrical Data	
	Diameter of Pipe D (mm)	Length of Pipe (m)
1–2 (vertical)	200	500
2–3 (horizontal)	150	700

Let us make the following calculations:

1. **Amount of Water.** This is given by the following formula:

$$Q_w = \frac{Q_k(\gamma_k - \gamma_m)}{\gamma_m - \gamma_w} \qquad (6.2)$$

$$\gamma_m = K\gamma_k + (1 - K)\gamma_w \qquad (6.3)$$

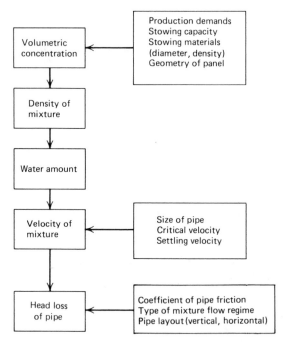

Figure 6.13 Design procedure of hydraulic stowing [2].

where Q_w = amount of water, in cubic meters per hour

Q_k = amount of stowing, in cubic meters per hour

γ_k = density of stowing material, in tonnes per cubic meter

γ_m = density of mixture (water + stowing), in tonnes per cubic meter

γ_w = density of water = 1 t/m³

K = volumetric concentration of mixture, the ratio of stowing v_k to stowing + water $(v_k + v_w)$

$$K = \frac{v_k}{v_c + v_w} = 0.3$$

$$Q_k = 100 \text{ m} \times 1.2 \text{ m} \times 2.0 \text{ m} \times 0.95$$

$$= 228 \text{ m}^3/\text{shift}$$

$$= \frac{228}{3} = 76 \text{ m}^3/\text{h}$$

The hydraulic stowing factor that 95% of voids can be filled up is 0.95.

$$\gamma_m = K\gamma_k + (1 - K)\gamma_w$$

$$= 0.30 \times 2.5 + (1 - 0.30)\,1$$

$$= 1.45 \text{ t/m}^3$$

$$Q_w = \frac{76(2.5 - 1.45)}{1.45 - 1}$$

$$\cong 178 \text{ m}^3/\text{h}$$

$$\frac{Q_k}{Q_w} = \frac{76}{178} = 0.42$$

2. **Velocity of Mixture**

$$Q = Q_k + Q_w = 76 + 178$$

$$= 254 \text{ m}^3/\text{h}$$

$$= \frac{254}{3600} = 0.070 \text{ m}^3/\text{s}$$

$$V_{1-2} = \frac{Q}{(\pi D^2/4)} = \frac{0.070}{[\pi(0.2)^2/4]}$$

$$= 2.22 \text{ m/s}$$

$$V_{2-3} = \frac{Q}{[\pi(0.15)^2/4]} = 3.96 \text{ m/s}$$

3. **Critical Velocity.** Durand [88] gives the critical velocity of a mixture to avoid settling as follows:

$$V_k = F_L \left(2gD\,\frac{\gamma_k - \gamma_w}{\gamma_w}\right)^{1/2} \qquad (6.4)$$

where V_k = critical velocity, in meters per second
F_L = coefficient, $d > 2$ mm, $F_L = 1.34$
g = acceleration of gravity, 9.81 m/s^2
D = diameter of pipe, in meters
γ_k = density of stowing material, in tonnes per cubic meter
γ_w = density of water = 1 t/m^3

Table 6.2 Critical Velocities in Pipes[a]

Pipe Diameter (mm)	Critical Velocitiy (m/s)
100	2.4
125	2.7
150	2.8
175	3.2
200	3.4

[a]See reference 2.

For sands (γ_k = 2.5) the critical velocity as just calculated is given in Table 6.2 for different pipe diameters. The velocity found in Table 6.2 for horizontal pipe is as follows:

$$V_{2-3} = 3.96 > V_k = 2.8 \text{ m/s} \qquad \text{(non deposit regime)}$$

4. **Friction Coefficient** (λ). In a turbulent flow the friction coefficients in a steel pipe can be taken as

$$\lambda_{1-2}: 200\text{-mm pipe} = 0.024$$

$$\lambda_{2-3}: 150\text{-mm pipe} = 0.025$$

5. **Friction Loss** (Head loss)

$$J_{1-2} = \lambda_{1-2} \frac{V_{1-2}^2}{2gD_{1-2}} \qquad (6.5)$$

$$J_{2-3} = \lambda_{2-3} \frac{V_{2-3}^2}{2gD_{2-3}} (1 + K\phi) \qquad (6.6)$$

$$\phi = 66 \left(\frac{\gamma_k}{\gamma_w} - 1 \right) \frac{gD_{2-3}}{V_{2-3}^2} \qquad (6.7)$$

where J_{1-2} = head loss in vertical pipe, in meters per meter of shaft (m/m)
λ_{1-2} = friction coefficient of pipe, 0.024
V_{1-2} = velocity in vertical pipe, 2.22 m/s
D_{1-2} = diameter of vertical pipe, 0.2 m
g = 9.81 m/s^2
J_{2-3} = head loss in horizontal pipe, in meters per meter

λ_{2-3} = friction coefficient of pipe, 0.025
D_{2-3} = diameter of horizontal pipe, 0.15 m
V_{2-3} = velocity in horizontal pipe, 3.96 m/s
K = concentration of mixture, 0.30
ϕ = Durand variable*
γ_k = density of stowing material, 2.5
γ_w = density of water, 1.0

$$\phi = 66 \left(\frac{2.5}{1} - 1\right) \left[\frac{9.81 \times 0.150}{(3.96)^2}\right]$$

$$= 9.28$$

$$J_{1-2} = 0.024 \frac{(2.22)^2}{2 \times 9.81 \times 0.2} = 0.030 \text{ m/m}$$

$$J_{2-3} = 0.025 \frac{(3.96)^2}{2 \times 9.81 \times 0.15} (1 + 0.30 \times 9.28)$$

$$= 0.50 \text{ m/m}$$

6. **Total Head Losses.** The losses of head per unit length are multiplied by the respective lengths to obtain the total losses as follows:

$$\Delta H = h_{1-2} \times J_{1-2} + l_{2-3} \times J_{2-3} \qquad (6.8)$$

$$= 500 \text{ m} \times 0.030 \text{ m/m} + 700 \text{ m} \times 0.50 \text{ m/m}$$

$$= 15 + 350 = 365 \text{ m}$$

---

*If $V_k < V \leqslant 17 \ V_s$, the slurry flow regime in the horizontal pipe is said to be "sliding bed flow" V_s is the terminal settling velocity of the particles. For rounded particle shape this value can be calculated by the equation

$$V_s = 0.55 \sqrt{d(\gamma_k - 1)}$$

For case above than,

$$V_s = 0.55 \sqrt{0.20(2.5 - 1)} = 0.3 \text{ m/s}$$

and

$$V_k = 2.8 \text{ m/s} < 3.96 < 17 \times 0.3 = 5.1 \text{ m/s}$$

In the sliding bed flow regime, Durand variable is given by Eq. (6.7).

As the length of shaft (vertical) is 500 m,

$$\Delta H < 500 \text{ m}$$

The mixture will reach its destination without any external source of pressure, like a pump. It can reach a horizontal length without any pump under the following conditions:

$$500 = 15 + l_{max} \times 0.5$$

$$l_{max} = 970 \text{ m}$$

Any distance further than 970 m from the shaft bottom requires extra pressure and the placement of a pump at the proper position.

6.4 ECONOMICS OF STOWING

The cost of pneumatic stowing and related cost analyses are given in Table 6.3 and schematized in Fig. 6.14 [2, p. 743], according to a study by Fettweiss [30] in the Aachen coalfield.

As shown in Fig. 6.14, when the mine operates on the caving

Table 6.3 Cost Analyses of Caving and Stowing Systems[a]

Cost Items	Commodity	Amount (tons)	Cost (NF/T)	Expenditure (NF/year)
Complete Caving System				
Cost of production in panel	Coal	840,000	1.81	1,520,000
Haulage in galleries and shaft	Rock	239,000	0.30	72,000
Surface transportation	Rock	239,000	0.20	48,000
Surface packing	Rock	624,000	0.73	455,000
Total				2,095,000
Stowing and Caving System				
Cost of production (stowing)	Coal	750,000	2.85	2,138,000
Cost of production (caving)	Coal	90,000	1.81	163,000
Haulage in shafts and galleries	Rock	624,000	0.42	261,000
Crushing plant	Rock	312,000	1.02	318,000
Total				2,880,000

[a]See references 2 and 35.

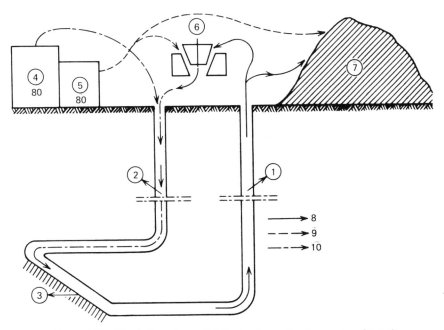

Figure 6.14 Circulation of materials in stowing and caving systems [2, 30].

system the yearly production is 840,000 tons of coal and with it 624,000 tons (430,000 m^3) of rock, which is put into a refuse pile (7). Of this rock, 239,000 tons result from development work and 73,000 tons from the washery (73,000 tons of +80 mm from screening, 312,000 tons of −80 mm from the washing process). When the mine operates on the stowing system, 750,000 tons of coal per year are produced by stowing and 90,000 tons are produced by the caving system, as there are not enough stowing materials available. In this system all the stone is used as the stowing material. The refuse of washery (312,000 tons directly and the rest, 230,000 tons mine rock plus 73,000 tons screening rock) is crushed and sent to the mine.

The amount of materials, cost of operations, and yearly total expenditures are summarized in Table 6.3 [30; 2, p. 744].

In the stowing system the combined amount of rock coming from the mine and rejects of the screening plant (−80 mm), 312,000 tons, has been crushed to −80 mm and mixed with the refuse of the washery, and a total of 624,000 tons is sent to the mine for stowing operations.

The cost of coal production per ton using the stowing system has risen from 1.85 NF to 2.85 NF (57% increase), while the entire expenditure of the mine has risen from 2,095,000 NF to 2,880,000 NF (a rise of 38%). Expenditures caused by subsidence and extra expenditure for roadway upkeep may reach this increase. So, in a populated area, the stowing systems are both economical and obligatory.

References

1. F. F. P. Kollmann and W. A. Cote, *Principles of Wood Science and Technology*, Vol. I, *Solid Wood*, Springer-Verlag, Berlin, 1968.
2. C. Birön and E. Arioğlu, *Madenlerde Tahkimat İsleri ve Tasarimi* (*Supporting and Design of Supports in Mines*), Birsen Kitabevi, Istanbul, 1980.
3. D. Beckett, and P. Marsh, *An Introduction to Structural Design: Timber*, Surrey University Press, London, 1974.
4. B. A. Jane, Mechanical properties of wood fiber, *Tech. Assoc. Pulp Paper Ind.*, **42**, 461–467 (1959).
5. F. F. P Kollmann, *Technologie des Holzer und der Holzwerkstuffe*, Vol. I 2nd. ed., Springer-Verlag, Berlin, 1951.
6. S. M. Dixon and M. A. Hogan, Tests on timber props, S.M.R.S. Paper No. 72 (1930).
7. N. C. Saxena and B. Singh, Props in longwall workings, *J. Mines, Metal, Fuels* (*India*), January (1969).
8. R. Baumann and C. Bach, *Elastizität und Festigkeit*, 9th ed., Springer-Verlag, Berlin, 1924.
9. H. Winter, *Richtlinen für den Holzflugzengbau*, Beitrage B II a 2, B II b, B II c, B II d, B III b, Berlin, 1944.
10. P. H. Sulzberger, The Effect of Temperature on the Strength Properties of Wood, Plywood and Glue Joints at Various Moisture Contents, S.C.I.R.O. Division of Forest Products, Melbourne, Australia, 1948.
11. F. E. Siimes, Mitteilungen uber die Untersuchung von Festigkeitseigenschaften der finnischen Schrittwaren, *Silvae Orbis*, (*Berlin*), No. 15, p. 60 (1944).
12. O. Graf, Tragfahigkeit der Bauholzer und der Holzverbindungen, *Mitt. Fachausschussber Holzfragen* (*Berlin*), **20**, (1938).
13. *Wood Handbook*, U.S. Dept. of Agriculture, Forest Service, Forest Products Laboratory, No. 72 (1955).
14. J. A. Nevlin and T. R. C. Wilson, The Relations of the Shrinkage and Strength Properties of Wood to its Specific Gravity, U.S. Dept. of Agriculture, Bull. No. 676 (1919).

15. R. Keylwerth, Spalten, Spaltbeanspruchnung und Querfestigheit des Holzes, *Holz Roh Werkst.*, **7**, 72–78 (1944/45).
16. Türkiye Köprü Ve Insaat Cemiyeti (Bridge and Construction Association of Turkey), Specifications on Wood Constructions, Ankara, 1958.
17. K. Szechy, *The Art of Tunneling*, 2nd ed. (English), Academiai Kiado, Budapest, 1973.
18. C. Birön, Undersea Coalmining and Its Application to Kozlu Mine (in Turkish), Habilitation thesis, İstanbul Technical University, 1962.
19. L. Siska, Problems Relating to Coal Extraction in Seams Containing Strong Sandstones in the Overlying Strata, Fifth International Strata Control Conference, London, 1972.
20. K. Terzaghi, *Theoretical Soil Machines*, 3rd Ed., Wiley, New York, 1965.
21. C. Birön and E. Arioğlu, Design of reinforced artificial roof of the thick lignite seam for the Soma mine of Turkey," *J. Mines, Metals, and Fuels*, 79–85, February (1978).
22. I. Evans, Face support requirements—A problem in "arching," *Int. Rock Mech. Min. Sci. Geomech.*, **14**, 1–5, Pergamon, London, 1977.
23. F. Spruth, *Steel Roadway Supports, Gluckauf Mining Handbook*, Vol. II, Verlag Glückauf, Essen, 1955.
24. H. F. Moore, *Materials Engineering*, McGraw-Hill, New York.
25. R. Peele, *Mining Engineers' Handbook*, Vol. II Wiley, New York, pp. 42–43.
26. National Coal Board Production Department, Trials with Sliding Arches, Information Bulletin 59/204 (1959).
27. E. Arioğlu, Factors Affecting the Design of Support Systems for Use in Roadways Associated with Longwall Faces in Coal Measure Strata, Ph.D. thesis, University of Newcastle upon Tyne, England, 1976.
28. R. V. Proctor and T. L. White, *Rock Tunneling with Steel Supports*, Youngstown Printing Co., Youngstown, Ohio, 1968.
29. S. S. Peng, *Coal Mine Ground Control*, Wiley, New York, 1978.
30. V. Vidal, *Exploitations des Mines*, Vol. 1, Dunod, Paris, 1961.
31. W. Goetze and W. Kammer, Die Aus Wirkungen von Strecken führung und Ausbautechnik auf die Querschnittsverminderung von Abbaustrecken, *Glückauf*, No. 15, August (1976).
32. R. M. Cox, A comparative evaluation of roof bolt anchors, *Monograph on Rock Mechanics Application in Mining*, Society of Mining Engineers, New York, 1977, Chapter 32.
33. E. Tincelin, P. Sinou, and O. Leonet, Soutenement Suspendu ou Boulonnage, *Rev. Ind. Miner.*, Special issue, September (1961).
34. K. Frecht, Grouting rock bolts using a perforated sleeve, *Tunnels and Tunneling*, October (1979).
35. Du Pont Co., Faslock Resin Anchored Bolt System, E. I. DuPont Co., Wilmington, Delaware, June (1974) 12 p.

36. J. M. Murphy, B. N. Whittaker, and M. J. Blades, Strata bolting: developments in design and application, *Colliery Guardian*, July (1972).

37. J. A. Franklin and P. F. Woodfield, Comparison of polyester resin and mechanical rock bolt anchor, *Trans. Inst. Min. Mech. Eng.*, July (1971).

38. S. Saluja and D. Singh, Recent Advances in Rock Bolting, Eighth World Mining Congress, Paper 5, Lima, Peru, 1974.

39. B. N. Whittaker and M. J. Blades, Resin Based Reinforcement of Rocks around Mining Excavations in Great Britain, Paper III-5, Symposium on Protection against Rock, Katowice, 1973.

40. D. F. Coates and T. S. Cochrane, Development of Design Specifications for Rock Bolting in Canadian Mines, Research Report R224, Mines Branch, Ottawa, 1970.

41. M. Dejean, The role of rock bolting and parameters in its selection, *World Constr.*, January (1977).

42. D. F. Coates and T. S. Cochrane, Rock Bolting, research and design specifications, *Can. Min. J.*, March (1971).

43. R. H Merill et al., Determining the Place Support in Roof with Rock Bolts, White Pipe Copper Mine, Michigan, U.S. Bureau of Mines, RI 5746, 1961.

44. S. D. Woodruff, *Methods of Workings in Coal and Metal Mines*, Vol. II, Pergamon, Oxford, 1966.

45. A. S. Bains, Experience with floor reinforcements at Birch Coppice Colliery, *Min. Eng.*, April (1978).

46. A. Ossacher, Kombination von Ankern mit anderen Ausbauelementen, Berg-und-Huttenmanische Monats hefte, 124, heft 6. Springer-Verlag, Vienna, 1979.

47. C. C. White, Mine Roof Support Systems, U.S. Patent No. 3427811, Washington, D.C., February 18, 1969.

48. C. C. White, Roof Support of Underground Mines and Openings, U.S. Patent No. 3505824, April 14, 1970 (continuation of Patent 3427811).

49. C. P. Mangelsdorf, Recent Progress in Roof Truss Technology, First Conference on Ground Control Problems in the Illinois Coal Basin, Carbondale, Illinois, August 22-24, 1979.

50. B. P. Sheorey, B. P. Verma, and P. Singh, An analysis of the roof truss, *J. Mines, Metals, Fuels*, August (1973) pp. 233-236.

51. E. A. Curth, Coal Mining Techniques in the Federal Republic of Germany—1971, U.S. Bureau of Mines, Information Circular 8645, 1974.

52. A. J. Barry, O. B. Nair, and J. S. Miller, Specifications for Selected Hydraulic Powered Roof Supports with a Method to Estimate Support Requirements for Longwalls, U.S. Bureau of Mines, Information Circular 8424, 1969.

53. T. E. Loxley, D. B. Lull, J. O. Rasor, Self-Advancing Roof Supports for

Longwall and Shortwall Mining, Final Report, U.S. Bureau of Mines, Contract No. 50144 128, April (1975).

54. D. C. Ashwin et al. Some fundamental aspects of face powered supports, *Min. Eng.*, August (1970).

55. B. N. Whittaker, A review of contribution made by powered roof supports on longwall mining, *Min. Eng. Dep. Mag.*, University of Nottingham (1975).

56. A. H. Wilson, Support load requirements on longwall faces, *Min. Eng.*, June (1975), pp. 479–491.

57. A. H. Wilson, Various aspects of longwall roof support, *Colliery Guardian Int.*, April (1978), pp. 50–55.

58. S. Sigott, "Die Belastungsmechanik des Strebausbau" Doktorat dissertation, Montanstischen Huchschule, Leoben, 1966.

59. J. P. Josien and C. Gouilloux, Present and future roof control and support in longwall faces in French coal mines, *Colliery Guardian Int.*, October (1978).

60. A. Bilinski and W. Konopko, Criteria of Choice and Use of Powered Supports, Symposium on Protection against Roof Falls, Paper IV-1, Katowice, 1973.

61. P. Stassen, Le Controle du Toit dans les Longues Tailles au Cours de ces 25 Dernieres Annees, Sixth International Strata Control Conference, Alberta, Canada, 1977.

62. J. J. Graham, Review of some recent powered support developments, *Min. Eng.*, June (1978), pp. 665–679.

63. C. Zillesen, Die Wirtschaftlichkeit des Schreitaubaus, *Glückauf*, March (1972).

64. C. Ford, An economic and technical appraisal of mechanized gate road drivage requirements, *Min. Eng.*, February (1970).

65. Turkish Standards Institute, Rules of Reinforced Concrete, T. S. 500, Ankara, 1971.

66. O. Neville, *Properties of Concrete*, Pitman, 2nd ed., London, 1973.

67. O. Graf, *Die Eigenschaften des Betons*, Springer-Verlag, Berlin, 1950.

68. F. D. Lydon, *Concrete Mix Designing*, Applied Science Publishers, London, 1972.

69. American Concrete Institute, Manuel of Concrete Inspection Publication, London, 1972.

70. American Concrete Institute, Shot-creting, Publication SP-14 Committee 506, Detroit, Michigan, 1966.

71. P. Strassen and H. van Duyse, The Driving of Large Section Crosscuts in Soft Ground and Lining with Concrete Blocks, Conference of Tunneling and Shafts, Institution of Mining Engineers, London, 1958.

72. K. Kondo, New Materials in Mining (Special Mining Method with Artificial Roof), Fifth World Mining Congress, F, Moscow, 1967.

73. C. Birön and E. Arioğlu, Design Principles and Specifications of the Reinforced Concrete Artificial Roof for the Gayeli Copper Mine of Turkey, III/18, II. The World Mining Congress, Belgrade, Yugoslavia, 1982.

74. Van L. Rabcewicz, The Neve Osterreichische Tunnelbauweisse, *Bauingenieur*, August (1965).

75. G. E. Wickham and H. R. Tieman, Research in Ground Support and its Evaluation for Coordination with System Analysis in Rigid Excavation, U.S. Bureau of Mines, Contract H0210038, 1972.

76. D. U. Deere et al., Design of Tunnel Liners and Support Systems, University of Illinois, for Office of High Speed Ground Transportation, U.S. Department of Transportation, Washington, D.C., 1969.

77. U.S. Army Corps of Engineers, Engineering and design of tunnels and shafts in rock, *Engineer Manual*, EM 1110-2-2901, Department of the Army, 1978.

78. S. A. Fjodorow, *Haugruben baue*, Veb Verlag Technik, Berlin, 1954.

79. F. Auld, Design of shaft linings, *Proc. Inst. Civ. Eng.*, Part 2, No. 67 September (1979) London.

80. G. Zahary and K. Unrug, Reinforced concrete as a shaft lining, *Proc. Eighth Canadian Rock. Mech. Symp.*, Toronto, December (1972).

81. C. Biron and E. Arioğlu, Design of Reinforced artificial roof for the thick lignite seam for the Soma Mine of Turkey, *J. Mines, Metals, and Fuels*, February (1978).

82. E. D. Crankshaw, In solid stowing on a mechanized face at Whancliffe Siltstone Colliery, *Iron Coal Rev.*, November (1960).

83. E. Rich, Rio Tinto's new pneumatic stower places waste fill efficiently, *Eng. Min. J.*, **158**, 84-87 (1957).

84. R. Farmin and C. E. Sparks, Sandfill method of dayrock resulted in these 12 benefits, *Eng. Min. J.*, **152**, September (1951).

85. M. A. Twindale, Backfill Methods in Canadian Mines, Information Circular 141, Department of Mines and Technical Surveys, Ottawa, 1962.

86. D. R. Corson, Stabilization of Hydraulic Backfill with Portland Cement, U.S. Bureau of Mines, Report of Investigation R. I. 7327, 1970.

87. P. Sarkka, Mines and Mining methods at Outokumpu Oy, Finland, *Min. Metall. Soc. (Finland)* (1979).

88. R. Durand and R. Gilbert, Transport hydraulique at refoulement des mixtures en conduites, *Ann. Ponts Chausees*, **130**, No. 3-4 (1960).

Author Index

Abrams, 176, 180
Adkins, ix
Agriculture Department of USA, 233
American Concrete Institute, 184, 236
Arioğlu, E., 70, 192, 198, 233, 234, 237
Arioğlu, N., ix
Ashwin, 236

Bach, 233
Bains, 235
Barry, 235
Baumann, 7, 233
Beckett, 233
Bilinski, 236
Birön, 193, 198, 202, 233, 234, 237
Blades, 100, 235
Bolomey, 176
Bridge and Construction Association of Turkey, 26
Bureau of Mines of USA, 90, 162

Coates, 107, 235
Cochrane, 107, 235
Corson, 233
Cote, 233
Cox, 91, 234
Crankshaw, 237
Curth, 235

Deere, 237
Dejans, 235
Dixon, 11, 233
Dowty, 139

Du Pont, 97, 234
Durand, 237

Ehrmann, 21
Evans, 38, 234
Everling, 31

Farmin, 237
Fettweiss, 230
Fjodorow, 237
Ford, 236
Franklin, 98, 99, 235
Frecht, 234

Gilbert, 237
Graf, 20, 176, 233, 236
Graham, 236
Goetze, 234
Gouilloux, 155, 236
Gullick, 139

Heinzmann, 68, 83
Hogan, 11, 233
Hutchinson, ix

Jane, 233
Josien, 155, 236

Kammer, 234
Keywerth, 234
Kollmann, 9, 11, 233
Kondo, 192, 236
Konopko, 236
Kunstler, 83

Leonet, 234
Loxley, 235
Lucas, v, ix
Lull, 235
Lydon, 236

Mangelsdorf, 119, 235
Marsh, 233
Merill, 235
Mikulak, ix
Miller, 235
Moll, 76
Moore, 234
Murphy, 99, 235

Nair, 235
National Coal Board of Britain, 166, 234
Neville, 236
Nevlin, 233
Newlin, 21

Ossacher, 235
Ostrava Research Institute of
  Czechoslovakia, 36, 127

Peele, 234
Peng, 98, 234
Protodyakonov, 28, 29

Rabcewicz, 197, 237
Rankin, 65
Resor, 235
Rich, 237
Riggan, ix

Sahin, ix
Saluja, 235
Sarkka, 237

Saxena, 12, 13, 233
Sheorey, 235
Sigott, 153, 236
Siimes, 17, 233
Singh, 12, 13, 235
Sinou, 93, 234
Siska, 34, 234
Smethurst, ix
Sparks, 237
Spruth, 234
Standards Institute of Turkey, 236
Stassen, 165, 236
Sulzberger, 233
Szechy, 234

Terzaghi, 38, 234
Tieman, 197, 237
Tincelin, 93, 234
Topuz, ix
Toussaint-Heinzmann, 68, 83
Twindale, 237

Van Duyse, 236
Vidal, 234

White, 119, 235
Whittaker, 99, 235, 236
Wickham, 197, 237
Wilson, 21, 152, 233, 236
Winter, 16, 233
Woodfield, 97, 99, 235
Wood Handbook, 233
Woodruff, 235

Yüksel, ix

Zahary, 237
Zillesen, 236

Subject Index

Additions to:
  gallery sets, 49
Advantages:
  of concrete, 171
  of powered supports, 165
  of roof bolting, 118
  of stowing, 211
Aggregates:
  of concrete, 177
  of granulometry, 173, 178
Allowable stress:
  in steel, 69
  in wood, 26
American standards:
  of concrete, 184, 205
  of I-beams, 67
  of steel, 64
American system, for powered supports,
  162
Analysis:
  of stowing costs, 230
  of stresses in articulated arches, 81
  of stresses in rigid arches, 70
  of wooden caps, 39
Anchorage capacity:
  of expansion-shell bolts, 92
  of resin bolts, 98
  of roof trusses, 122
  of slot-and-wedge bolts, 91
Application:
  of concrete in mines, 184
  of powered supports, 167
  of roof bolts, 111
Articulated:

arches, 76
caps, 125, 130
Artificial roof:
  concrete, 192
  design, 202
Austrian system, for powered supports,
  153

Base of powered supports, 144
Bearing plates of props, 128
Bending strength:
  of rocks, 30
  of steel, 63
  of timber, 13, 21
Brinnel hardness of steel, 65
British standards:
  of cement, 203
  of concrete, 183
Brushings of gateways, 211
Buckling factor, wooden sets, 44
Buckling strength of timber, 12

Canopy:
  of caps, 127
  of powered supports, 144
Cap design:
  articulated, 76
  timber, 41
Carbon in steel, 63
Carrying capacity:
  of powered supports, 149, 154, 155, 157
  of roof bolts, 106
Cellulose, 4
Cement:

in concrete, 173
materials, 222
Characteristic curve:
  of friction props, 129
  of hydraulic props, 132
Characteristics:
  of Clement profile, 69
  of concrete, 175
  of H-beams, 66
  of rail profile, 69
  of rocks, 30
  of roof bolts, 103
  of steel, 62
  of supporting elements, 65
  of T-H profiles, 69
Chocks, wooden, 32
Chock type powered support,
  139
Clement profile, 69
Compactness of concrete, 177
Concrete:
  Abram's cone, 180
  advantages of, 171
  aggregates, 173
  application, 184
  artificial roof, 192
  blocks, 188
  cement content, 173
  compactness, 177
  curing conditions, 178
  design, 192
  dosage, 173
  engineering characteristics, 175
  fluid, 180
  flying ash, 174
  granulometry, 181
  importance of, 171
  making, 180, 195
  making, numerical example, 195
  maintenance, 182
  monolitic, 188
  pipeline, 182
  plastic, 179
  pouring, 182

  preparation, 192
  shaft lining, 189
    numerical example, 200
  slump, 182
  strength, 183
  supports, 171
  transportation, 181
  water/cement ratio, 175
  water content, 175
  wet, 179
  working conditions, 177
Cone of Abram, 180
Consolidated material:
  stowing, 222
  strength, 224
Control system, powered supports, 146
Convergence:
  in powered support face, 165
  in powered supports, 148, 155
  in prop-and-cap face, 165
Conveyor, snaking, 127
Cost analysis, stowing, 230
Critical velocity in hydraulic stowing,
  228
Crushing strength of timber, 8
Curing time:
  of concrete, 178
  of resins, 97
Cut-and-fill stopes, with roof bolts, 117
Cutting-loading machine, 127

Deformation of steel, 62
Description:
  of powered supports, 144
  of roof bolts, 90
  of yield arches, 82
Design:
  of additions to gallery sets, 49
  of artificial roofs, 202
  of caps, 55
  of concrete, 192
  of concrete shaft lining, 198
  of friction props, 135
  of hydraulic stowing, 223

of longwall cap, 56
of longwall supports, 55
of moll arches, 76
of optimum size, 55
of powered supports:
  American system, 162
  Austrian system, 153
  distance of, 147
  French system, 155
  German system, 149
  load density, 148
  Polish system, 157
  yield pressure, 147
of props-and-caps, 134
of rigid arches, 72
of roof bolting, 109
of roof bolts, 105
of roof trusses, 120
of shaft indention, 201
of shaft lining, 198
of shotcreting, 197
of side posts, gallery set, 43
of steel arches, 70
of steel caps, 136
of wooden caps, 41, 46
of wooden gallery set, 39
of wooden wedges, 44
Dimensions:
  of powered supports, 145
  of roof trusses, 123
  of yield arches, 84
DIN specification:
  of H-beams, 21541, 66
  of steel, 21544, 64
Disadvantages:
  of concrete, 172
  of powered supports, 166
  of stowing, 213
Dosage of concrete, 173
Du Pont resin cartridge, 97

Economics of stowing, 230
Effect of carbon in steel, 63
Efficiency factor of props, 134

Elasticity modulus of steel, 64
Elastic limit of steel, 63
Elongation of steel, 63
Engineering characteristics:
  of concrete, 175
  of steel, 62, 78
  of timber, 4
English system of powered support design,
  150
Estimation of yield arches, 86
Everling pressure formula, 31
Evolution of longwall supports, 125
Expansion-shell bolts, 92

Face, prop-free, 125
Face conditions, 139
Factor:
  of buckling, wooden sets, 44, 48
  of efficiency, friction props, 134
  of expansion, longwall, 33
  of loading (Everling), 31
False roof, 32
Fatigue, timber, 20
Faulted area:
  longwall, 116
  roof bolts, 118
Fibers, 4
Fibrous structure, 4
Floor:
  heaving, bolting, 115
  intrusion, 48, 136
Fluid concrete, 179
Flying ash in concrete, 174
Frame type of powered supports, 139
French system, for powered support
  design, 155
Friction prop:
  bearing plates, 128
  characteristic curve, 129
  description, 126
  floor intrusion, 136
  instantaneous, loading type, 129
  locking systems, 129
  numerical example, 135

profiles, 131
prop efficiency, 134
slowly sinking type, 129

Gateway roof bolting, 112
German system, for powered support
 design, 149
Glocken yield arches, 83
Graf formula, 176
Granulometry:
 of aggregates, 178
 of concrete, 181
 of shotcreting, 187
Gravity stowing, 215
Grouted roof bolt, 95
Gunite, 184

Hand stowing, 213
Hardness of steel, 65
Head losses, hydraulic stowing,
 229
Heaving of floor, bolting, 115
Hydraulic:
 critical relocity, 228
 chock, 139
 head losses, 229
 power supply, 146
 props, 126, 131
 stowing, 219
 stowing design, 225

Immediate roof, 32
Importance of concrete, 171
Installations:
 of powered supports, 169
 of resin roof bolts, 99
Instantaneous loading, 130
Intrusion to floor:
 gallery wooden sets, 48
 friction props, 136

Jacks of powered supports, 146

Kunstler yield arches, 83

Legs of powered supports, 146
Length of roof bolts, 107
Lignin, 4
Load:
 density of powered supports, 148
 extension relations of roof bolts, 103
 factor (Everling), 31
 on gallery set, 28, 31
 on powered supports, 152
Locking systems, of friction props, 129
Longwalls:
 roof bolts, 114
 steel cap design, 136
 steel prop design, 134
 steel supports, 125
 wooden cap design, 55
 wooden post design, 58

Machines, pneumatic stowing, 215
Maintenance of concrete, 182
Making of concrete, 180
Material:
 cementing, 222
 steel, 61
 stowing, 209
 wood, 4
Mechanical:
 characteristics of steel, 62
 properties of resins, 97
 properties of wood, 23
 stowing, 215
Metal mines, roof bolting, 116
Middlepost design, 49
Minimum cap diameter, 54
Modulus of elasticity, steel, 64
Moisture effect, wood, 21
Moll arches, 80
Moments of inertia of beams, 65
Moments in rigid arches, 75
Monolitic concrete, 188
Moving platform, shaft lining, 191

Normal loads, 75
Numerical examples of design:

anchorage of roof bolts, 92
concrete making, 195
concrete shaft lining, 200
friction props, 135
hydraulic stowing, 225
roof bolting, 109
roof trusses, 121
shaft indention, 201
shotcreting, 197
tightening of roof bolts, 94, 107
wooden gallery caps, 46
wooden gallery posts, 47
wooden longwall, 58

Operating data of powered supports, 145
Optimum design of gallery set, 52
Optimum distance of caps, 55
Optimum size of caps, 55

Packing strip, 213
Pectin, 4
Penetration to floor, 48
Pipeline for concrete, 182
Pipes, pneumatic stowing, 215
Plastic concrete, 179
Pneumatic:
  stowing, 215
  stowing pipes, 218
Poisson's number, rocks, 30
Polish system, for powered support
    design, 155
Porosity of rocks, 30
Post, design, 43
Pouring of concrete, 182
Powered supports, 167
  advantages, 165
  applicability, 167
  bases, 144
  canopies, 144
  carrying capacity, 149, 153, 154, 155, 157
  chock type, 135
  comparisons, 167
  control system, 146
  convergence, 155, 165

description, 144
design, 147-164
development, 138
dimensions, 145
disadvantages, 166
frame type, 139
hydraulic power, 146
installations, 169
jacks, 146
legs, 146
pressure formulas, 32
principles, 138
roof conditions, 160
roof indices, 159, 162
shield type, 139, 142
skid, 144
yield pressure, 146
Preparation of concrete, 192
Pressure:
  on longwalls, 32
  on roadways, 27, 31
  on wooden supports, 27-39
  steps in calculation, 27
Principle:
  of friction prop, 128
  of hydraulic prop, 131
  of powered supports, 138
  of roof bolting, 89
  of roof trusses, 119
  of yield arches, 82
Profiles:
  of friction props, 131
  of steel beams, 65
Prop:
  and-cap face, 162
  density, 134
  free face, 125
  friction, 126
  hydraulic, 126
  numerical example of design, 135
Protodyakonov pressure formula,
    27

Quarries for stowing materials, 211

Rail characteristics, 69
Rankin ratio of profiles, 69
Ratio:
  of Rankin, 69
  of water/cement, 175
Refuse piles, 211
Rejects, concentrators, 211
Resin:
  bolt anchorage, 101
  curing time, 97
  mechanical properties, 98
  roof bolts, 96
Rigid arches:
  design, 72
  moments, 75
  normal loads, 75
  numerical example, 74
Rocks:
  physical characteristics, 30
  strengths, 30
Roof:
  conditions, 160
  false, 32
  immediate, 32
  indices for powered supports, 159, 162
  pressure, 31
Roof bolts:
  advantages, 118
  anchorage:
    expansion-shell, 93
    numerical examples, 92
    slot-and-wedge, 92
  anchorage capacities, 104
  application, 111
  carrying capacity, 106
  cut-and-fill stopes, 117
  design of, 105
  diameter, 108
  expansion-shell type, 92
  for faults, 118
  for floor heaving, 115
  grouted type, 95
  load-extension relations, 103
  longwalls, 112, 114

  metal mines, 116
  numerical example, 109
  principle, 89
  resin type, 96
  room-and-pillar workings, 112
  slot-and-wedge type, 90
  spacing, 107
  for spalling, 117
  stability, 105
  testing, 101
  tightening, 93
  varieties, 90
  wooden type, 101
Roof trusses:
  anchorage, 122
  design, 120
  dimensions, 123
  numerical example of design, 121
  principles, 119
  statics, 120
Room-and-pillar bolting, 112

Section modulus of beams, 65
Servo wedge, 130
Shaft:
  indention design, 201
  lining design, 189
  numerical example of design, 201
Shear strength:
  rocks, 30
  timber, 21
  verification, 42
Shield supports, 122, 139, 142
Shotcreting:
  design, 195
  granulometry, 187
  numerical example, 197
Skid, powered support, 144
Slenderness:
  number, side post, 44
  ratio of timber, 12
Slot-and-wedge type roof bolt, 90
Slowly sinking friction prop, 129
Slump of concrete, 182

Snaking conveyor, 127
Source of stowing materials, 210
Spacing of roof bolts, 107
Spalling, roof bolts, 117
Specifications:
 steel, 64
 wood, 26
Stability of bolted rock, 105
Standards:
 cement:
  American, 205
  British, 203
 concrete:
  American, 184
  British, 183
  Turkish, 183
 H-beams DIN 21541, 66
 I-beams, American, 67
Static analysis:
 rigid arch, 70
 roof truss, 120
Steel:
 allowable stress, 69
 American specifications, 67
 arches, 76
  particulated caps, 125, 130
 beams, 65
 characteristics, 61
 DIN specifications, 64
 effect of carbon, 63
 elastic limit, 63
 elongation, 63
 engineering characteristics, 62
 hardness, 64
 H-beams, 65
 hydraulic props, 131
 importance, 61
 longwall supports, 125
 material, 61
 mechanical characteristics, 62
 mode of failure, 64
 moments of inertia, 65
 prop, 125
  friction type, 128

 hydraulic, 131
 prop design, 134
 properties, 64
 Section modulus, 65
 specifications, 64
 stress-deformation, 62
 structure, 62
 tension strength, 63
 working scheme, 132
 yield arches, 82
 yield point, 64
Stowing:
 advantages, 212
 consolidated, 222
 disadvantages, 213
 economics, 230
 gravity, 215
 hand, 213
 hydraulic, 219
 hydraulic design, 223
 importance, 209
 machines, 215
 mechanical, 215
 materials, 209
 numerical example of design, 225
 pneumatic, 215
 systems, 213
Strength:
 concrete, 183
 consolidated stowing, 224
 rocks, 30
 steel, 63
 working steel, 64
Stress:
 allowable, steel, 69
 analysis rigid arches, 70
 deformation of steel, 62
Strip packing, 213
Structure of steel, 62

Tensile strength:
 rocks, 30
 roof bolts, 101
 steel, 63

timber, 5
Testing of roof bolts, 101
Tightening:
  expansion type roof bolts, 93
  force, numerical example, 107
Timber:
  advantages, 3
  allowable strength, 22, 26
  audio indications, 15
  bending strength, 13
  buckling strength, 12
  compressive strength, 10
  crushing strength, 8
  defects, 5
  disadvantages, 3
  effects of fatigue, 20
  effects of grain, 9
  effects of knots, 17
  engineering characteristics, 4
  factor of completeness, 16
  fibers, 4
  forms of break, 17
  knots and cracks, 5
  layers of age, 4
  material, 3
  mechanical properties, 23
  moisture/content, 11
  natural defects, 6
  safe strength, 22
  shear strength, 21
  strength, numerical values, 21
  support design, 39-59
  tensile strength, 5
  torsional strength, 25
Toussaint-Heinzmann profiles, 68
Transportation of concrete, 181
Turkish standards of concrete, 183

Ultimate strength of steel, 64
Unit weight of steel beams, 65
U-Section yield arches, 83

Varieties of roof bolts, 90
Velocity, critical, hydraulic stowing, 228
V-Section of yields arches, 83

Walking supports, 127
Washery rejects, 211
Water/cement:
  ratio for concrete, 175
  ratio formulas, 176
Wedges, wooden, design, 44
Wedge type of roof bolts, 90
Weight of steel beams, 65
Wet concrete, 179
Wood:
  classes, 26
  macroscopic structure, 4
  mechanical properties, 23
  roof bolts, 101
  supports, design, 39-59
  see also Timber
Working:
  conditions of concrete, 177
  hydraulic props, 132
  strength of steel, 64
  yield arches, 82

Yield arches:
  estimations, 86
  numerical example, 86
  point of steel, 64
  pressure of powered supports, 147
  valve pressure, 146
Young's modulus of rocks, 30